OUR DROWNING WORLD

By the same author:

NOISE POLLUTION
LONDON'S DROWNING
FLOODSHOCK

OUR DROWNING WORLD

Population, Pollution and Future Weather

Antony Milne

INSTITUTION OF ENVIRONMENTAL SCIENCES

PRISM PRESS

Published in Great Britain in 1988 by

PRISM PRESS
2 South Street
Bridport
Dorset DT6 3NQ

and distributed in the USA by

Avery Publishing Group Inc.
350 Thorens Avenue
Garden City Park
New York 11040

Copyright © Antony Milne 1988

ISBN 1 85327 004 0

All rights reserved. No part of this book, text, illustrations or other material may be reproduced in any form without written permission from the publishers.

Typeset by Maggie Spooner Typesetting, London
Printed and bound in Great Britain by
Biddles Ltd, Guildford and King's Lynn

CONTENTS

Introduction   1

PART ONE: THE URBAN HOTHOUSE
Chapter One: The Exploding Megacities   10
Chapter Two: The Urban Heat Bomb   22
Chapter Three: The Hot Skies   32
Chapter Four: The Balding Earth   45
Chapter Five: The Greenhouse Effect   58

PART TWO: THE COMING FLOODWAVE
Chapter Six: The Warming Warning   72
Chapter Seven: The Cosmic Connection   87
Chapter Eight: The Disintegration of Antarctica   100
Chapter Nine: The Floodwave Effect   111
Chapter Ten: Floodshock!   129

Epilogue: Who Wins, Who Loses?   140

Postscript   144

Bibliography   147

The future is not what is used to be — Arthur C. Clarke

For my mother and her progeny

INTRODUCTION

May 8th 1984 was a prestigious and important day for London, a day commemorated with a Royal flotilla that made its majestic way along the Thames from Westminster to Woolwich Reach. But it was a historic day for other, more profound, reasons. For the official opening of the biggest movable flood barrier in the world was a tacit recognition of the powerful geologic forces that had become an insidious menace to Londoners, whose city was sinking into its bed of clay at the rate of over a foot a century.

London has for decades been suffering from a land subsidence problem, coupled with the menace of a destructively high 'surge' tide. But now earth scientists are quietly expressing their fears that the English Channel and the North Sea itself could rise by several more feet within just 200 years from now. Without the new barrier it is a certainty that London would be rapidly submerged under sea water well before the next century was over.

The projected rise in global sea levels within the next few centuries means that already civilization is in peril. Scientists now admit that the melting of the polar ice sheets could raise ocean levels by as much as 100 feet, inundating all the world's major cities situated near coastlines, and drowning sizeable regions of agricultural land. Indeed, the slipping, melting ice packs have now become, like the giant asteroid that threatens annually to strike the Earth, one of the greatest natural Doomsday dangers facing us. In particular the giant Antarctic ice sheet, two miles thick and larger than Europe, is so tenuously grounded to land that it could literally break away at any moment and create tidal waves around the world.

These dangers arise because the Earth is getting warmer. Both the US National Academy of Sciences and the Environmental Protection Agency have recently published alarming and detailed warnings to this effect, and have hinted that rising tidal levels will be *the* environmental problem of the coming century. Indeed the timescale involved, in the light of what is known about the excruciating slowness of geologic time, will be remarkably brief. World temperatures, say the EPA, would probably rise nearly four degrees F above what they are now by the year 2040, and go up a further 5F by the end of the 21st century — a total warming even greater than that which abruptly brought about an end to the last Ice Age.

The main causative factor in this warming is the growth of thermal pollution around the globe. Much of this extra heat, in turn, arises from increasing amounts of carbon dioxide gas that is being pumped into the atmosphere. This gas allows the Sun's shorter ultraviolet wavelengths and visible radiation to warm the Earth. But at the same time it absorbs the longer infrared wavelengths the surface tries to radiate back into the atmosphere. So the heat builds up, as it does in a greenhouse.

The main culprit, conclude the atmospheric scientists, is the burning of coal and other fossil fuels. These carbon-rich energy sources release an estimated 5½ billion tons of carbon dioxide (a colourless, odourless gas) into the world's atmosphere annually. Hence the activities of Man figure prominently in this coming crisis. The speed of the forthcoming melting of the polar glaciers will be determined by the rapid warming of the upper atmosphere, and it is now recognized that this heat-up has taken place largely since 1950 when the world's cities started to expand quite rapidly.

Quite apart from this, massive geophysical forces are bringing about similar dangers. Intermittent Ice Ages and widespread flooding have occurred repeatedly throughout history, and it is a geologic certainty that such processes will continue. What is not certain is the exact time span over which these dramas have been played out.

In the past disputes raged about the pace of geologic time. Some — the Catastrophists — believed that Earth's past had been forged by violent upheavals and celestial bombardments. Their opponents instead believed that the features of the planet — the mountains, rivers and ravines — had been created painfully slowly over billions of years by the slow attrition of the wind, Sun and rain.

The worldwide flood legends of ancient times were for centuries explained in catastrophic terms following the tradition established by the 17th century writings of Gottfriend Leibniz and John Woodward. Many later scientists, when they came to interpret Earth's lithologic history, simply borrowed this catastrophic viewpoint. But when James Hutton published his revolutionary *The Theory of the Earth* in 1785, Catastrophism took the first of many intellectual batterings.

And yet the legacy of Hutton today serves to remind modern geologists that a sense of history is important. The past informs us about the future. With geology, as with other sciences, tracing the record of past developments aids the formation of 'dynamic' models that are often made to operate in a vacuum.

However, renewed speculation about dramatically rising sea levels and high tides suggests that the Catastrophists have finally joined ranks with the Uniformitarians after years of bitter conflict. The pace of events, it seems, has quickened. In addition the palaeontologist and the climatologist — new breeds of postwar scientist — have had a vital influence on present-day geology. So indeed have the glaciologists working in Antarctica, and the atmospheric scientists at the Moana Loa volcano in Hawaii. Earth science today, like Cosmology, is not only the fastest developing and exciting academic discipline, it is also the most complex and demanding.

Experts are hence modest about their predictions. The National Academy of Sciences, while concerned about the magnitude of the possible climatic change, admit that their predictions lack the rigour of a mathematical science. Scientists from the University of East Anglia's Climatology Unit, on returning from an EEC project on the greenhouse effect in November 1983, admitted that prediction was going to be difficult. The mechanism of weather and climate are so complex that few are even certain that rising temperatures in the Northern Hemisphere will cause an increase in rainfall in that region.

The impact of the Sun, of course, is of vital significance in any discussion of climate. And with the rapid progress in astrophysics in recent years we now know much more about how the Sun behaves. We know more about its past and its likely future. However none of this knowledge is comforting, since it only reinforces the fear that the future of every living thing on this planet is at the mercy of the tiny loops, spots and flares that frequently appear on the surface of the Sun.

What is disturbing is the new knowledge coming to light which shows that the polar regions seem to be disproportionately affected by solar fluctuations. In any event even small changes in the balance of ultraviolet and infrared radiation can cause distorted repercussions in the convective processes that arise in the form of drifting warm air masses circulating the globe. For instance a two percent reduction in the Sun's heat could lower Earth temperatures by about 5C, enough to bring on a new Ice Age within 100 years from now. Astrophysicists fear that the Sun may be returning to behaviour patterns which were 'normal' in the past, to bring devastation to Earth's biosphere, soon to make Man as extinct as the famous dinosaurs.

But what if the Sun's radiation *increased* by two or more percent? And what if this was added to the anthropogenic Greenhouse warming? The climatologist, however, is not likely to rush to judgement. The most important question confronting him today is to what extent sea-level changes are caused by climatic change, and to what degree they are (like, perhaps, the waxing and waning of the Ice Ages) cyclical and therefore predictable. He may ask whether there is any evidence that Ice Ages and Water Ages occur at roughly 20,000 year intervals, as the Earth changes its angle of axis relative to the Sun every 40,000 years or so.

The geologist, on the other hand, seeks answers closer to home. Nowadays he would be obliged to integrate an environmental perspective into his research because of the recognition of a highly varied range of complex inputs that can produce quite different phenomena. Landslips, storms, soil erosion, atmospheric warming, all have an unavoidable human dimension. According to George Woodwell, director of the Marine Biological Laboratory in Massachusetts, vast areas of the biosphere — which includes the ozone layer — are being threatened by radiation from nuclear plants, the growing aridity of the land surfaces, and the general toxification of the environment.

Us human beings, although we can only introduce minute inputs into the biosphere, seem — like the minuscule variations in the Sun's radiation — to have an inordinate effect on things. For example a loss of topsoil of only about one percent a year would spell disaster for much of the world's agricultural lands within a short space of time. Factors such as these may explain the inexplicably fleeting nature of civilizations. Many eminent authors, like Arnold Toynbee, have highlighted the often malign role that both climatic vicissitudes and land abuse have had on the human race.

And history shows, too, that as a prelude to the decline of a civilization the climate became intensely disturbed. Entire races became squeezed in the vicious pincer movements of encroaching frost from the north, and expanding deserts to the south, or — if they lived on the coast — had to flee from an ever onrushing tide.

Present generations are similarly threatened. The evidence today is that the world's weather systems are being disturbed not only for the reasons that the Environmental Protection Agency have suggested, but because of the imbalance between atmospheric moisture and terrestrial water, and because of the growing amounts of dust and sulphur particles ejected from volcanoes.

And the evidence is growing more empirical by the day. As with the rapid advance in plate tectonic theory in the 1960s — spurred by the first generation of modern computers — so study of the climate is now made with the help of computer models of the atmosphere.

And it is this greater depth of analytical and mathematical power in climate analysis that shows how weird and unexpected is the impact of shifting trends in the weather. When for example simulated carbon dioxide concentrations in the atmosphere are made to reach twice present values, a projected rise in temperature of about 3C occurs. But the putative impact of such a rise is far from uniform. At any given latitude, the effects could differ dramatically. Earth could undergo a rapid warming with distinct pockets of local cooling.

Indeed, recent weather aberrations, with floods, drought, heatwaves and storms occurring indiscriminately, could well be evidence that existing carbon dioxide levels are already, ahead of schedule, fulfilling the computer models with depressing accuracy. Existing predictions point to an alarming shift in the main climatic zones towards the poles. This would benefit regions like the Sahel which would get more of the equatorial rains, but at the same time it would warm the poles by from three to five times the average. The desert regions would also be pushed northwards, possibly to adversely affect agriculture in the Mediterranean region, which would suffer a marked cooling. It would also disrupt the agricultural economy of Russia's virgin lands, already threatened by drought.

What is becoming clear, however, is that the term 'micro climate' will enter more and more into the vocabulary of the meteorologist as time passes. Micro climates, referring to weather syndromes peculiar to cities and dense urban areas, must inevitably figure more prominently in future theories about long-term climatic change.

The micro climate will, for the first time, ensure that the scientists' *law of aphasy* becomes operative. This is because the climatic environment is said to cause effects which out-distance any natural or organic adaptation to such effects. So it will be the law of aphasy that will dictate the iron logic of the runaway Greenhouse Effect. It predicts that myriad micro climates, by exponential growth and by aggregation, will assume the characteristics of the macro climate. And by the middle of the next century it may become warmer than it has been for 8,000 years. And the century after that a 70 million-year climatic optimum may be reached. But long before that crop production in the world's main arable regions would have been seriously disrupted. America's CIA has even warned of economic instability, civil unrest and war as a result.

But the greatest danger facing Earth, as we have seen, arises from what happens to the global ice sheets. At present there are some eight million cubic miles of ice in the world's polar regions, the vast majority of it in the southern hemisphere. This ice tends to melt a little during summer and refreezes during winter. But this balance, maintained for the last 7,000 years or so, will shortly no longer be in equilibrium.

The important point about the ice sheets is that a vicious circle soon arises when a melting, no matter how slight, begins. Since the white ice reflects more sunlight than bare earth, it follows that as the ice begins to shrink the Earth absorbs more solar radiation, which melts the ice faster and faster, and so on. Soon, all the ice packs will melt into the ground and run to the oceans, which are already overflowing, in the sense that sea levels encroach beyond the littoral shelves well into existing shorelines.

Indeed, even now many parts of the arctic regions are ice free several months of the year longer than in earlier years. The great birch forests of Canada have also suffered severely from the attacks of insect pests that have been spurred northwards by the warmer weather. Tropical fish have been reportedly caught off Long Island, and shipping in the Hudson Bay is active a month longer than in the period before the Second World War. According to Richard F. Flint of Yale University the entire North American continent is rising as the land recovers from the weight of the last Ice Age. This probably accounts for the present height of the Great Lakes and for the ever higher tides around Long Island as the melting ice surges southwards. Flint's theories, that we are still at the tail-end of the last Ice Age, are adhered to by many other scientists.

Hence, after a brief intermission, Man may now be on the verge of re-living the perennial human drama of retreating from an onrushing tide. This book will attempt to explain why and how the future warming of the Earth will cause the giant ice packs still remaining to melt, and the likely impact this will have on Man and his environment.

PART ONE:

THE URBAN HOTHOUSE

Chapter One:

THE EXPLODING MEGACITIES

Somewhere in the world in July 1986 an Asian woman gave birth to a baby, and the world's population for the first time topped the 5,000 million mark.

This calculation was arrived at by an army of statisticians and demographers working for the United Nations, and who were aided by the big computer at the Population Institute in Washington DC. The institute's director, Werner Fornos, admits no doubts. He and his team regularly process data made available from the various UN agencies and from national censuses, plus estimates from the 156 countries where the UN has representatives.

The computers calculated that some 10 million Chinese and more than 13 million Indians were, or were about to be, born in 1986. These two countries are among the largest and poorest of the world. China, worried about rampant population growth, set out in the mid 1970s to reduce its annual birth rate, then standing at 34 for every 1,000 inhabitants. In less than a decade it has cut that figure to 20 — a remarkable achievement, but one that was largely due to the ruthless enforcement of a state decree forbidding Chinese married couples from having more than one child. Now China's birth rate, like that of the West, is less than one percent a year.

Indeed, couples across the world are having fewer babies, and United Nations predictions have twice been revised downwards to a growth rate of 1.73% per year. The US Census Bureau projects the drop in the live birth rate between 1975 and 2,000 AD to go from 30 to 26 per 1,000, although the African rate, already high at 47 per 1,000, will decline to only 38.5.

And yet the human family continues to grow inexorably. Between now and 2110, according to the United Nations Fund for

Population Activities, regional populations will increase as follows:

South Asia	1.4 billion to 4.1 billion
East Asia	1.2 billion to 1.7 billion
Africa	400 million to 2.1 billion
Latin America	400 million to 1.2 billion
Europe	450 million to 500 million
USSR	265 million to 380 million
North America	248 million to 320 million
Oceania	23 million to 41 million

Source: Omni Future Almanac, Sidgwick and Jackson, 1983

The UNFPA puts the world stabilization figure at around the year 2110 when the global population will have expanded to some 10,500 million (10.5 billion). But there is no sign as yet of any *reduction* in the world's population. The global economy, its ecosphere, its natural resources will continue to be under immense strain, and there will be a myriad problems concerned with massive urbanization, and poverty. Unemployment, one of the most divisive and disturbing issues of the 1980s, will be far worse than it is now, with the population only half as great.

As Jyoti Singh, the UNFPA's director of information, says, the world has become complacent and shortsighted. Even China cannot escape from the sheer weight of its existing population. China still produces 10 million extra babies a year. Growth rates are still high in the rest of the world. Africa continues to give especial concern, as population proliferates at well over 2% a year, and almost half the inhabitants are still under the age of 15. True, India over the last 20 years or so has made strenuous gains in achieving self-sufficiency in food, but infant mortality is still high and the average life-expectancy is only 53 years. The World Bank reckons that India's total population will reach not less than 1.6 billion in the next couple of generations. And the world's next billion, alarmingly, is due at just before the end of *this* century. Any newspaper article celebrating 'the end of the World Baby Boom' would be grossly misleading.

And, as encouraging as stabilization of population might be, we still have 124 years of remorseless growth ahead of us with apparently no prospect of any longer-term reductions after that point has been reached. France, Italy, West Germany and Britain are all set to retain high similar populations of between 55 and 60 million in perpetuity, although their optimum populations ought to be half the existing numbers. It is worth remembering that Sweden, with a similar amount of fertile living space and the same cultural attributes, has only a very small population and probably the best quality of life in the world. Britain, with slightly less space than other major European nations, has 911 people per square mile, and is already the most urbanized nation in the world. In fact Britain is four times more crowded than China.

In the meantime progress even towards the goal of stabilization will be a long and arduous one. Global population rates will only begin to slowly reduce when birth rates fall below the replacement level of 1.8 children per family, which will then bring the growth rate down to nought percent from its current level of 1.63%, which itself is down from the 1969 peak of 2.1%. It is worth remembering that still, every year, there are 50 million *more* mouths to feed.

Things are bound to get worse as time passes. As a further example of unjustified optimism about population trends let us take the marginal improvement in Mexico City's fertility rate. In 1975 the city's population was expected to double within 15 years, whereas the doubling time has now been advanced to 27 years. But this still means that the city will 'house' 32 million by the year 2011. Most of these inhabitants, unless there is some miracle of planning or the sudden tapping of enriching resources, will live in rain-soaked shanties. Mexico City's air is already so polluted it clogs the lungs with the daily equivalent of the sooty gases from 40 cigarettes. The city produces 6,000 more tons of rubbish daily than it can collect.

What is happening to Mexico City is not just a reminder of population growth, but of massive *urban* growth. Current rates of fertility will be just as high in the rural areas, but there will be much more migration into the big cities, which will soon become Megacities, bursting at the seams. In Africa two out of five people will, through rural-urban drift or birthright, be domiciled in urban areas.

This is why even the quality of life in the West, where the total number of inhabitants will increase by less than half a percentage point a year, will decline as the disbenefits of urban living are spread further throughout the community, and additional strains on local economies and social services will make themselves felt.

There is already frequent crime, drug abuse and homelessness. There is much unemployment in what were once thriving northern cities in Britain, France and Germany. There, the heavy metal-based industries (shipbuilding, steel and car manufacturing) have shrunk drastically as the post-industrial society (a term frequently dismissed as a sociologist's cliché) actually emerges, requiring different kinds of skills and greatly reduced — and widely dispersed — levels of staffing.

Even in America large, prosperous, cities like Dallas have growing numbers of welfare recipients and homeless people. In the last 20 years refugees from the mill towns of the rust belt and the small farming communities made bankrupt through declining subsidies have set up new, but infinitely poorer, communities along the sprawling ringroads surrounding the Sunbelt cities of America.

In the meantime some of the difficulties confronting Third World governments are in danger of becoming insurmountable. Many of the world's future poor will be concentrated in cities far too large, and with very little of the basic services we take for granted in the West. Even if Bangladesh, for example, succeeds in reducing its population growth to 1.7%, it will still hold 357 million people in giant urban shanty towns within 65 years. Then there will be 10 times as many people on every square mile of land as there is in crowded Britain today.

Within that time India, too, will become the world's most populous country, with 1.5 billion, because of its still high fertility rate (4.7 per woman, compared with America's 1.8). Calcutta, which has great difficulty housing its present total of 11 million, will be even harder pressed to cope when the total reaches the predicted 16.6 million in 2025.

Even higher rates of fertility are to be found in black Africa, where some national rates have actually increased in recent years. Africa as a whole only had a population of 175 million in 1940. Compare this with the 513 million achieved in 1983, and the anticipated 877 million by the end of the century. Kenya's fertility rate of 8.1 live births must make it the fastest growing nation in the world. Every day there are 1,637 new Kenyans born. At this rate the population could increase to 83 million in 2025 from its present level of about 20 million. Says John Cleland, past co-ordinator for the World Fertility Survey in Kenya: 'Already Kenya's development programme is running at a standstill — just to keep up. Unless something is done to check population growth, Kenya will simply crack under the strain'.

Hence high fertility is at the root of the problem. Cultural pressures play their ignoble part. Women have a low status in Africa, and there has recently been a breakdown of traditional patterns of birth spacing. Many peasant men have more than one wife, and refrain from sex during the breast-feeding period, only to demand their marital rights later when it is no longer 'safe'. Hostility to anti-natalist policies by Kenya's middle-ranking officials doesn't help: family planning is often portrayed as a subtle form of genocide.

This lax attitude by Third World governments to rising levels of fertility, if unchecked, will mean that by about 2050 the vast majority of people, about 7 billion, will be living in the poorer regions of the globe. There will be a disturbing preponderance of the young, sickly, illiterate and politically volatile. Inevitably Nature will soon step in to impose its own arbitrary methods of population limitation. Not all babies born in impoverished countries will survive into adulthood. There is chronic malnutrition in Africa, which has the highest rate of infant mortality in the Third World. In India four million infants died in childhood in 1984, out of the 23 million that were born. A further 16 million will suffer permanent physical and mental damage. Of those Indian babies born in 1984 only three million will develop into fit and robust adults.

The Flight to the Cities
Over 3,000 years ago, the world's largest cities were to be found in Biblical lands — along the great river valleys of the Tigris and Euphrates and the Nile. Thebes, for example, had a population of roughly 100,000, and Babylon had 54,000. Later Rome became the world's biggest city with almost 600,000 inhabitants. By AD 1000 Constantinople had nearly half a million. It was not until the late 18th century that Peking housed a million people — the first city to do so.

Later it was industrialization that spurred the rapid flight to the cities. England, of course industrialized first, and London's population reached 2.3 million in 1850. But just 50 years later the population had already reached an astonishing 6.5 million. The infrastructure of London — its roads and thoroughfares — was laid down in the 19th century.

World urban growth got under way on a massive scale after 1950, with already large cities doubling in size up to the year 1975. The rapid flight to the cities, as the population reaches ever higher figures, is spurred on by the search for greater employment and life chances. By the year 2000, according to UNFPA projections, over 50% of the world's entire population will be housed in cities, that is triple the number of metropolitan inhabitants in 1950. By then some 630 million young adults will be attempting to join the labour force of the poorer countries. And, as labour-saving technology will become cheaper for industrial managers to instal, the struggle to gain employment will intensify. Africa will set the pace among the Third World countries, with a 981% increase in city dwellers predicted between the years 1950 and 2,000. Asia will be second, with a 652% increase by the century's end, raising the total to 790 million.

Writing shortly after a major UN sponsored population conference at Oxford University in 1974, Peter Wilsher and Rosemary Righter reported that the size of the world's cities has been growing dramatically. Taking a communal aggregation of 100,000 as representing a city, they suggested that by the turn of the century the balance between town and country will be the opposite of what it was in 1960; i.e. there will be between 6,000 and 7,500 million people to accommodate. Two-thirds of them will be in ramshackle urban areas. The 'great wens' of India, Nigeria, Brazil and Indonesia will have outstripped the Parises, Londons, and New Yorks by a factor of around two and a half to one.

How big can the cities get? According to the United Nations big cities start at half a million. After the million-cities come the multi-million-cities, containing five million upwards. The super-conurbations, like Tokyo-Yokohama and Greater New York, weigh in at 12.5 million upwards. At present the largest urban area in the world is Tokyo, with 25 million. It has kept well apace of London, which was second to New York after World War II. But by the year 2,000 London will not even be in the top 25. And Tokyo will pale into insignificance.

From now on we must start talking about Megacities, and they will be found almost entirely in Asia and Latin America, especially the latter. But the new urban agglomerations of the future will offer a standard of life vastly inferior to that experienced by the present London or Tokyo. They will be sprawling, festering places of habitation, especially at the edges. It is at the fly-blown outskirts of Istanbul, Karachi, Rangoon and Rio that the real city explosion is taking place.

Already, by 1980, there were 10 Megacities with populations exceeding 10 million, whereas 20 years previously there were only three — New York, London and Tokyo; significantly all 'western' cities. By the year 2,000 the Third World will be well into the Megacity League, with as many as 25, including Cairo, Calcutta, Sao Paulo and Rio de Janeiro. Latin America, in fact, is burgeoning with Megacities: Buenos Aires, Lima and Santiago must now be added to a growing list. Between 1945-65 Caracas quintupled its population, and doubled it again during the next 20 years. Most Argentinians now live in urban agglomerations.

As recently as 1950 there would only have been a short-list of million-cities. But by 1975 there were 191 entries, with new names like Curtiba in Brazil and Lyallpur in Pakistan. The total of urban dwellers in 1975 had already risen to 515 million. Since then the imbalance between north and south, developed and underdeveloped, countries has grown. It is the smaller number of cities in the latter that is contributing substantially more to the urban population explosion.

The cities of Asia, too, are expanding at a staggering rate. And not simply in the poorer regions. Japan, a rich and technologically advanced country, housed only 18% of its people in urban areas in 1920. By 1940 this figure had grown to 40%. Now the figure has leapt to 75%.

UN demographers project a grand total of 82 cities in the four-million range, nearly two-thirds of them in the poorer regions. Two out of three large cities across the globe in varying million-range categories will be in the less developed regions. By 2110 almost 90% of the world's inhabitants will be living in Third World regions, with the more industrialized nations, with populations rising at a much slower rate, facing a decline in their share of the world population, from 24% in 1980 to 13% in 2110. By then the white races of the world will be in a distinct minority, and more and more of the leaders of western countries will have to deal with Third World problems.

MILLION-CITIES IN THE WORLD
Cities of one million or more inhabitants

Source: United Nations

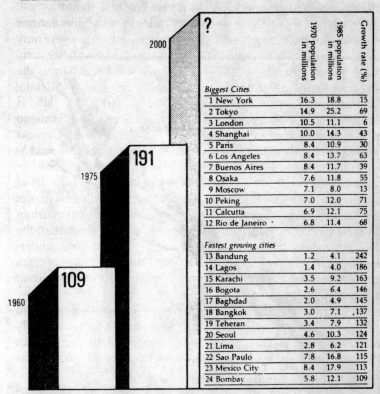

	1970 population in millions	1985 population in millions	Growth rate (%)
Biggest Cities			
1 New York	16.3	18.8	15
2 Tokyo	14.9	25.2	69
3 London	10.5	11.1	6
4 Shanghai	10.0	14.3	43
5 Paris	8.4	10.9	30
6 Los Angeles	8.4	13.7	63
7 Buenos Aires	8.4	11.7	39
8 Osaka	7.6	11.8	55
9 Moscow	7.1	8.0	13
10 Peking	7.0	12.0	71
11 Calcutta	6.9	12.1	75
12 Rio de Janeiro	6.8	11.4	68
Fastest growing cities			
13 Bandung	1.2	4.1	242
14 Lagos	1.4	4.0	186
15 Karachi	3.5	9.2	163
16 Bogota	2.6	6.4	146
17 Baghdad	2.0	4.9	145
18 Bangkok	3.0	7.1	137
19 Teheran	3.4	7.9	132
20 Seoul	4.6	10.3	124
21 Lima	2.8	6.2	121
22 Sao Paulo	7.8	16.8	115
23 Mexico City	8.4	17.9	113
24 Bombay	5.8	12.1	109

The Emerging Megalopolis

We must now come to terms with the *Megalopolis*, another word to describe a Megacity that has virtually joined up with another. Rather like a business amalgamation, the Megalopolis is highly functional. Its existence is determined by favourable transport costs and trade barriers across narrow routes, to the mutual benefit of expanding manufacturing areas.

At present there are about 20 of these sprawling urban areas in the world. They range up to 60 miles in size, with Boston–New York–Washington as the oft-quoted example. Jean Gottman, the French geographer, in his mammoth 1961 study of urbanized north-east America, referred to the 'Bosnywash'. This was an emergent continuous string of central cities, suburbs and satellite areas. Even then it covered 450 miles, stretching along the old US Highway No 1 running along the eastern seaboard of the United States. Today it includes Philadelphia, Newark, Baltimore and Worcester — a total of 53,000 square miles in which live some 90 million people.

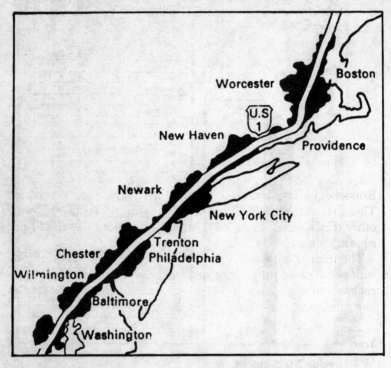

The Bosnywash east coast Megalopolis of the United States
Source: The International Urban Crisis, Thomas Blair, 1979, Hart-Davis

Some urban sociologists point to a similar giant metropolitan region in Britain encompassing London, Luton, Northampton and Birmingham — a total of 118 miles. Thomas L. Blair, in his book 'The International Urban Crisis', goes one step further and talks of the Golden Triangle of Europe, which includes not only Birmingham and London but spans the Channel to include Paris,

The Exploding Megacities

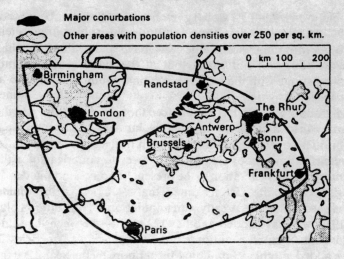

The Golden Triangle of Europe
Source: Thomas Blair, ibid.

Brussels, the Randstad, the Rhine-Ruhr region, and Frankfurt. This massive built-up zone includes major conurbations and other dense urban areas with human densities of over 250 per square kilometre.

The frightening implications of the Megalopolis are that cities with populations running between tens and hundreds of millions each will soon be commonplace. Wilsher and Righter envisage the 'metropolitanized' world shaping up as follows:

	City Populations		(Unit: million)
Year	100,000+	500,000+	1 million+
1985	1,371–1,522	954–1,048	789– 883
1990	1,596–1,838	1,119–1,283	939–1,106
1995	1,851–2,220	1,307–1,570	1,112–1,387
2000	2,131–2,681	1,520–1,922	1,309–1,738

Source: The Exploding Cities, Wilsher & Righter, Deutsch, 1975

It is quite clear, then, that a massive demographic transition is getting under way at a galloping pace. The UN has estimated that as early as the year 2,000, the urban proportion of the globe will house 3,234 million, with the balance of 3,260 in the rural sector. Within a generation from now the world's urban population will double, even if 2.0–1.7% global growth rate is wildly out, because of the relentless, and accelerating, rural-urban drift. Not all cities will become Megalopolises. Some will have that fate postponed because they will not spread out at the same geometric rate commensurate with the increase in human numbers: people will be packed in tighter.

Migration, in short, has unbalanced the scheme of things. In the Third World it adds an intolerable burden to the municipal authorities already struggling with high indigenous fertility rates. Whereas the influx of European refugees to America since the founding of the Federation to the present has only amounted to 50 million or so, the UN reckoned that 24 cities in the underdeveloped countries, already with populations of one million plus in 1950, will have added another 122 million between then and 1985. And 51 similar million-cities in Europe and America will have added another 93 million. Furthermore by the year 2,000 the developed regions of the world will have more than trebled. Europe's urban population will have grown by 70% by then.

It is the longitudinal dimension that is so worrying about population growth. The rate of change is disturbingly fast. Eric McGraw, Director of Population Concern, a major British population control pressure group, paints a chilling historical profile of what exponential growth actually means.

From the 18th century onwards the major factor involved has not been changes in the birthrate, but changes in the death rate. By 1721 western Europe was free of the dreadful plague that hitherto had been a highly effective restraint to population growth. It took from the beginning of time until about 1830 for the human family to reach its first thousand million. It took only a further 100 years to gain a second thousand million; the third in 30; the fourth in 15. The population 'took off', says McGraw, from the 18th century, when revolutions in agriculture, industry and public hygiene got under way.

Man is now in an acute moral dilemma. Medical advances are constantly raising the life expectancy of infants, thus unwittingly vitiating whatever gains may be had from population control policies. On the other hand the size of the human family would be even bigger if every one of the million children conceived each day survived the womb; as we have seen about half are aborted naturally because of the poor health and malnutrition of the mother.

The Exploding Megacities

The sociological implications are clear. The World Bank estimates that by 2025 there could be widespread overcrowding in shanty towns, plus hunger and joblessness, and unmanageable urban growth. Politically the outcome could be more violence in the streets, with the familiar vicious cycle of more authoritarian government precipitating more terrorism, and so on. There could be heightened global instability.

But more importantly there is bound to be more environmental devastation and soil erosion, especially if a world population approaching 10 billion has to be fed. As this book will soon make painfully clear the total impact that Man and his Megacities will have on the environment will soon be having a disastrous and irreversible impact on our climate.

Chapter Two:

THE URBAN HEAT BOMB

Man, by himself, is an energy converting machine. He converts food, as fuel, into motion. And according to a basic principle of science we call the Second Law of Thermodynamics the energy that is so produced is eventually, and inexorably, degraded into heat. Just sitting quietly a human being gives off 450 British Thermal Units (BTUs) of heat per hour, which is enough to warm one pound of water to one degree F.

But man has also invented vastly more powerful machines that are also energy converters on a prodigious scale; like automobiles. A large family saloon driven at 70 mph gives off 750,000 BTUs. In 1925 total energy consumption and conversion had already reached 44 quadrillion BTUs, and, growing at a rate of 5% annually since then, world energy use leapt to 345 quadrillion BTUs in the 1980s. Science — and logic — decrees that the world must be getting warmer.

The population explosion has for years been deplored for 'environmental' reasons. But the devastating impact the sheer size of the human family is having on the global climate has only recently been appreciated. We have built ourselves massive heat and moisture emitting cities across a great part of the world's land surface. And they are all in danger of generating more waste heat than they receive from the Sun. Already Chicago and Washington have frost-free growing seasons a month longer than the surrounding rural areas.

A great danger facing mankind, and the biosphere, is the additional heat that is being pumped into the environment at a rate faster than can be re-radiated back into space. And, like the positive feedback features of exponential human and biological growth, the more the temperature of the Earth rises the more terrestrial surface characteristics will change (i.e. polar ice sheets will shrink, aridity and desertification will increase), and so the cycle is perpetuated. It matters little that the original stimulus is measured only in tiny percentages, since the knock-on effect is disproportionately huge. Some experts feel that if we add only as much as 2% globally to the value of incoming radiation, the entire geosphere will be threatened.

But this threat looms larger every year. Across the urbanized world human energy conversion rates equal about 1/15,000 of the absorbed solar energy. And at the present 5% growth rate in energy consumption the global heat balance will match one-thousandth that of the Sun, and this will bring about noticeable changes in Earth's climate. This, according to one Russian scientist, will be the ecological upper limit.

Throughout history Man has been his own enemy. Modern technology has given rise to highly destructive wars, and we will now for ever face the threat of total annihilation in a nuclear holocaust. But scientific prowess has also promoted both the longevity and wellbeing of the human race. We have allowed our numbers to increase phenomenally to such an extent that we are capable of melting the polar ice caps just by sitting in our offices and homes. We are a threat to the entire global ecology just by being here.

The threat became a potential reality as early as 1650 AD, when power and technology for the first time made their impact on human social evolution. The population stood at 500 million. It was then that population growth and the application of science to lighten the workload and to shape the environment took off together. In spite of periodic recesssions and episodes of negative growth, the long term trend has been remorselessly upward. From 1812 onwards early industrial societies were beginning to deplete irreplaceable fossil fuels as they started to use energy derived from coal and oil in increasing quantities. After Newcomen's invention of the steam engine in 1705 more and more factory work was done by machines rather than by people or animal energy or water.

People and fuel-burning machines all suddenly began to multiply with alarming alacrity. A short while later electricity was discovered. Electricity is the juice of technology — clean, silent, powerful; producing vast material gains, the ultimate in technological achievement. It now energises virtually every collective human activity, and becomes all the more precious as automation and computerization advances to take over much of what was human drudgery.

The power blackout in eastern parts of the US in November 1965 was yet final proof that advanced industrial societies are almost totally dependent on continuous supplies of power. That the blackout occurred in America was salutory: Americans are now said to be consuming up to 40% of the world's energy requirements, while constituting only 6% of the world's population. Every single American man, woman and child is said to use the energy equivalent of the bodily labour of 500 slaves.

Power consumption is expected, globally, to claim 50% of all available energy by the year 2,000. Just one high-rise building, whether in Dallas, Sao Paulo or London, often utilizes more electricity for its essential functions than that needed by many villages, or even a small city.

Vast amounts of fossil energy are required for the world's packaging industries. Aluminium soft drink and beer cans, glass bottles and polypropylene bottles use up the most energy. According to Richard North, in his book *The Real Cost*, the world manufactures 60 million tonnes of plastics annually — a growing ingredient in modern packaging — and in the US alone synthetic material production consumes about 3% of the country's oil and natural gas.

However, one widely quoted study performed by the American Institute of Energy Analysis actually predicts a steady rate of consumption of energy (with zero growth) in America of 125 quads up to the year 2025 (1 quad is 1 billion million BTUs). Growth in western Europe of commercial energy consumption is reckoned to be 2%, with 4% in eastern Europe, and 1.5% in the Third World. The steady state in America is thought to be largely due to stringent self-imposed conservation measures, and to the fact that modern US domestic appliances ar enow some 50% more energy efficient than those built 10 years ago.

The future world thermal energy problem, as with population growth, is likely to be concentrated in the developing world, still energy-hungry and unlikely to adopt conservation techniques at least until industrial 'take-off' has arrived. The problem at the moment is not commercial fuel consumption so much as the energy-wasteful slash-and-burn techniques of peasant landholders.

Fossil fuel burning may be five times its current level by the year 2200. In Chapter 4 we will examine the harm to the global climate this factor alone is likely to cause.

The Heat Island Effect

Luke Howard, a chemist, was probably the first to discover — as early as 1818 — what is now known as the *heat island effect*. Howard published a series of volumes on London's urban climate, in which he noted, even in the early 19th century, that night 'is 3.70 degrees warmer, and by day 0.34 degrees cooler, in the city than in the country'.

Maximum heat-island effect of 11 US Metropolitan areas. (1978 figures)

City	Additional temps. reached over prevailing rural temps. ($C°$)	Population (millions)
Louisville, KY	6.5	0.89
Baltimore, MD	5.2	2.14
Washington, DC	5.2	3.02
Cincinnati, OH	5.1	1.38
Indianapolis, IN	4.5	1.14
Dayton, OH	4.5	0.84
St Louis, MO	4.4	2.37
Richmond, VA	3.8	0.57
Columbus, OH	3.3	1.07
Kansas City, MO	3.2	1.30
Petersburg, VA	2.6	0.04

Now other scientists, like Helmut Landsberg, a senior climatologist at the University of Maryland, recognize that the chief characteristics of urban zones are the heat and high pressure weather systems they create, and which are often quite absent in the surrounding rural localities.

The reason for this phenomenon is explained with the aid of some simple physics. Different surface materials have varying levels of heat conductivity. Spongy, frequently moist, arable soils generally have a low rate of heat conductivity, and a higher rate of light reflectivity. In addition the superstructure of city buildings and pavements, the tarmacadamed roads and concrete walls, act

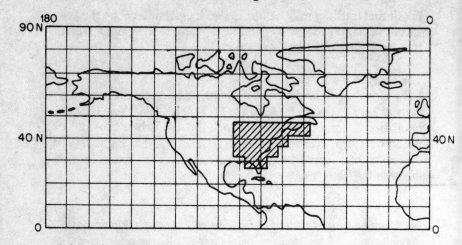

Enhanced computerized images based on satellite surveys clearly shows the reality of thermal pollution in the eastern United States.

Source: Landsberg, Helmut E., The Urban Climate, Academic Press, (US) 1981

as a giant thermostat, soaking up the sunlight and retaining the heat of the day — both solar and anthropogenic heat.

Simple micro-scale studies soon proved what was happening to built-up areas. An experiment was done on a five-storey brick building with a paved court measuring 32 × 42m, next to which was a wide grass lawn. Landsberg reported that even two hours after sunset the courtyard was some 5C warmer than the air temperature, and 1C higher than the surrounding grassland.

Let us take the case of Tokyo, which, since 1920, has been experiencing temperatures rising way above the regional trend. There was a downward dip in this spiral during the 1940s, when the city was largely destroyed in the War. But following an extraordinary period of reconstruction from 1946 to 1963 urban temperatures rose nearly 1C. Other rapidly expanding Japanese towns were matched by scientists with smaller towns which had minimal rates of growth. The results were similar to the Tokyo experience: an almost perfect statistical relationship between city size and heat emission.

Similar records were obtained for Paris. The city is now about 2C warmer at its centre than out of town. A lengthy monitoring exercise covering 67 years up to 1968 showed that in 1891 the city

The Urban Heat Bomb

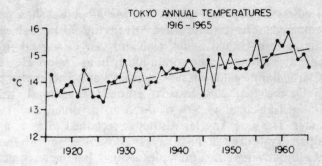

Long term temperature trends in Tokyo

Temperature rises in Japanese cities 1936-65

Rapid growth		Slow (or no) growth	
City	Temp Rise (C°)	City	Temp rise (C°)
Tokyo	0.032	Nemuro	0.005
Osaka	0.030	Tyoshi	0.011
Kyoto	0.032	Hikore	0.020

temperature was 1C warmer than the surrounding environs, rising to nearly 2C by 1968. Even earlier studies of Paris's heat island effect have come to light. One Jean-Dominique Cassini is said to have installed a thermometer in a cellar built deep below the Paris Observatory. As early as 1671 regular checks were made of the reading, which varied between 11.7C and 11.9C up until about 1870. Then a steady rise in the cellar's temperature occurred until it reached, in 1969, 13.5C.

Other recent observations have emerged from weather stations at airports and rural localities. Jean Detweller, a French meteorologist, found that during the decade 1951 to 1960 isotherm readings at the airports showed temperatures of 10.6C to 10.9C, compared with the centre of Paris, where it was 12.3C.

What was reasoned scientifically and measured thermometrically was proved recently with the aid of enhanced satellite images taken from the air. These infrared images show that even the outskirts of metropolitan areas soon become incorporated into the heat island. The famous Kew Gardens, for example, with large parklands situated some 13 miles from the centre of London, was viewed as part of a giant island largely centred on the closely built-up surrounding neighbourhoods.

Needless to say the rise of the Megalopolises has greatly accentuated the heat island effect. Already the 'Bosnywash' region of the eastern United States seaboard can be viewed from a satellite in earth orbit as a continuously lit up region. Heat from this particular megalopolis now equals 5% of the net energy of the Sun. By the year 2050, experts estimate, Bosnywash will be emitting heat equal to 50% of the Sun in winter, and 15% in summer. The Los Angeles district is reckoned to have a heat dissipation rate of 6% of the solar energy received, and this fraction is steadily growing. Other examples abound. Baltimore, containing two million plus people, is a distinct 2C warmer than Woodstock, Maryland, some 25 km away.

Centigrade increases in summer temperatures in selected US cities over that prevailing in rural areas.

City	Period of record	Warming rate (C°)
Cleveland, OH	1895–1941	0.03
Boston, MA	1895–1933	0.026
Washington, DC	1895–1954	0.024
Tampa, FL	1895–1931	0.016
Baltimore, MD	1894–1954	0.02

There is another disturbing characteristic of the urban heat island. Big cities, it appears, can also interfere with both the circulation of local airflows and the moisture balance. This came as a surprise to scientists. City air was generally considered to be drier simply because it contained less rain or dew-soaked vegetation. And as rainwater tended to run-off down sewers instead of permeating into soil, it was believed there would be a smaller transevaporation cycle. But the scientists left man and his machines out of the picture. Indeed, there is more than enough extra humidity in cities, largely arising from exhaust fumes and human respiration (remember each human being consists largely of water!).

Another phenomenon peculiar to the city is the obstacle effect. The jagged roughness of the urban skyline can stir up the weather. Rain-making processes in the atmosphere can be stopped in their tracks, the rain clouds then grow bigger and darker, and precipitation is more violent. The effect is more marked at the end of warm, sultry summer days. There is for example a higher

Urban effects on summer rainfall in nine US cities.

City	Effect observed	Maximum change (mm)	(%)
St Louis	Increase	41	15
Chicago	Increase	51	17
Cleveland	Increase	64	27
Indianapolis	Indeterminate	—	—
Washington	Increase	28	9
Houston	Increase	18	9
New Orleans	Increase	46	10
Tulsa	None	—	—
Detroit	Increase	20	25

frequency of midday cloudiness at La Guardia Airport in New York City compared with the more outlying Kennedy Airport.

Italian meteorologists cite the example of Turin, Italy. The city grew from 700,000 to 1.2 million from 1952 to 1969, with a rapid road building programme to keep pace with the spiralling number of cars. There was a distinct increase in light, but frequent, showers. In the rural regions adjacent to Turin the effect of industrial aerosols also caused an increase in rainfall. Other Italian scientists discovered a 17% increase in precipitation in Naples for the period 1946–75.

The Warming Cooling Towers

As the insatiable human demand for more energy rises, the ratio of heat to power grows relentlessly. The population bomb necessitates the continual building of additional, or larger, power plants. There are now more desalination plants. The fossil fuels — oil, coal and gas — that these massive generators feed off are typically 40% efficient, creating four units of electricity for every 10 units of carboniferous energy they consume. But as the world increasingly turns to nuclear powered plants, this low level of efficiency drops still further to 30%, so more heat escapes into the environment via the cooling system.

The simplest type of system is that known as once-through cooling. Water is simply sucked in from a local reservoir, pumped round the conduits of the generating station, and then discharged back into another reservoir some 5C to 8C warmer. The problem is that not all power stations are situated so conveniently next to abundant water supplies. The cooling tower is thus a popular alternative to once-through cooling, where cold air is sucked into the plant, and waste heat ejected through a chimney tower.

The danger to Earth's ecosystem, then, is that energy is being transformed into heat at a much faster rate than would occur naturally. The city has one serious drawback: it lacks the genetically coded stability and resilience to be found in lakes, prairies and forests.

A man-made world lacks the kind the homeostasis inherent in natural ecosystems. A river, for example, will quickly recover from a momentary infusion of poisons. However, the interrelated human activities that rely on a city's services are not organic in the strictly biological sense, so a sewage strike, for example, brings dislocation to city life. Moreover the factors that brought the city into being in the first place — population growth and energy consuming technology — are essentially those that are at the root of the environmental crisis. In other words the city is a biological abnormality, and its growth akin to pathological, rather than ecological, processes.

But it is a pathology, unfortunately, that is highly resistant to treatment. The positive feedback, the mutual reinforcement, is the fate of organisms in disequilibrium. High city temperatures, paradoxically, demand the consumption of yet more energy. The warmer it is the more air conditioners are used, and the more electricity is consumed. Atmospheric pollution control, and noise and tetraethyl emission controls on cars, mean that yet more energy is released into the environment. So the attempts to control other forms of pollution succeed only in increasing thermal pollution.

We have seen that each human being gives off 450 BTUs of heat, and that world energy use now equals 345 quadrillion BTUs, taking into account all other energy-converting sources. According to Vsevolod Troitsky, a member of the Soviet Academy of Sciences, current world energy production approaches 10^{13} watts. This provides roughly two kilowatts per head of population, with about 10 kilowatts consumed by each inhabitant of the West.

Dr. Troitsky says that Nature has set an upper limit on Earth's population at just double the present figure, that is just below the stabilization figure of 10.5 billion, projected by UNFPA for 2110 AD. The ecological 'safe limit', he says, would be for man-made energy to reach roughly 10^{14} watts, ten times what it is now, or one-thousandth of the total solar energy reaching Earth.

Unfortunately Dr. Troitsky tells only half the story. If the safe limit for energy inputs alone will arrive in the year 2110 then we are already in trouble, simply because, as we shall see, other atmospheric changes are taking place which are compounding the waste heat problem. In the meantime, regardless of whether the impact of urban heat is significant in global terms at present, the *magnitude* of the microclimate effect (i.e. measured as a phenomenon by itself) is severe. It is only a matter of time before regional, and ultimately global, climatic variations become apparent to all. And it will be in the polar regions where the impact will hit the hardest.

But there are other ways in which cities are continuing to warm the world, and we must now consider these before examining the ways in which climate is likely to be influenced.

Chapter Three

THE HOT SKIES

One might, in recent years, be forgiven for assuming that 'environmental problems' were diminishing in importance, as other problems crowd the global stage. Even environmentalists think there have been some signs of improvement in doomsday trends in the 15 or so years since The Environment became a media buzz word. The flyleaf of a new book on ecology even has the temerity to declare that 'population growth rates are slackening spectacularly everywhere, except Africa'. The truth, as we have seen, is rather different.

There is one curious anomaly about the contemporary world that has led to such erroneous conclusions. The West is now heading fast towards the much talked about 'post industrial society', with the commensurate decline of the conventional polluting smokestack industries. The rise of the Sunrise microchip technologies and the growing importance of the service sector of the economy suggest to many that the pollution problem is on the way out for good. Certainly the air of western cities is becoming cleaner, but only in the strictly optical sense. Instead, as this chapter will make clear, the nature of the invisible anthropogenic gases now being pumped into the atmosphere poses an even greater threat to the Biosphere than the black smoke belched out from the vanishing industrial chimneys of yesteryear.

In any event, the microchip-based utilizing sectors of modern economies are just as heavily dependent on the use of electricity, and hence fossil fuels (nuclear reactors are still comparatively rare). Cities are getting larger, and more spread out. And there are more cars on the road, and more people are consuming more of the world's resources. Man-made dust and pollution is still posing disbenefits and creating massive disamenities. According to US climatologist Murray Mitchell, latterly of the US Weather Bureau, we collectively pump about 30% of all minute particles less than five microns across into the atmosphere every year, and they remain suspended for about nine days before falling to Earth.

The ecosystem, although amazingly resilient, is still vulnerable to overload, to over-exploitation, and to the excessive emissions of wastes. The major disbenefit of pollution is not only toxicity but volume. Even biological degradable waste will cause pollution if it enters the ecosystem too rapidly to be processed, in the same way that excessive rainfall causes floods. In other words we are faced, in most cases, with a purely quantitative problem.

There is mounting evidence, too, that society is becoming overly dependent on synthetic substances. Soaps made from natural fats are now replaced by plastics and detergents. Synthetic fibres now replace natural ones. But the by-products created in their manufacture can be dangerous and insidious — various acids, phenols, nitrates and chlorinated hydrocarbons.

Indeed, the threat from the indiscriminate use of chemicals is already lethal. Dioxin is a new and deadly chemical that has caused deaths and injury to many people. Today as many as 1,000 new chemical formulas come on to the market every 18 months. Around 35,000 to 50,000 chemical compounds now available in US pharmacies have been classified by the US Environmental Protection Agency as either definitely or potentially hazardous to human health. Many of these have long-term toxic effects, as they cannot be broken down naturally, and may still be around for hundreds, even thousands, of years.

Herein lies the threat to the atmosphere. Most scientists now estimate that pollution amounts to some 300m tonnes at any one time. This represents about 25% of the total natural number of contaminants (such as volcanic dust, wind-blown dust and soil particles, methane and sulphur dioxide).

The global reach of pollution is now immense, and many parts of the Third World are still experiencing the smog-laden skies common in postwar Europe and America. A recent World Health Organization study shows, for example, that cities where fresh air is becoming virtually non-existent include Cairo, Rio and Delhi. Dozens of others exceed the health targets set by the WHO, and among those cities with the most unhealthy air are Milan, Madrid and Tehran. Many cities were way over the top of the WHO's recommended limits for air grubbiness — Tehran at 222 micrograms per cubic metre, Bogota (120), Cairo (105), Madrid (196) and Havana (101). For gravimetric methods measuring other airborne particles the figures were as much as 50% higher — Lahore in Pakistan had a phenomenal 690 micrograms.

Northern Europe, predictably, gets a cleaner bill of health. In June 1985 a report published by the Organization for Economic Cooperation & Development (OECD) claims that carbon monoxide levels in its member countries were somewhat diminished, and that there was reduced pollution from certain insidious chemicals, such as DDT, polychlorinated biphenyls (PCBs) and mercury compounds. There were fewer tanker accidents and oil spills, plus better protection of national parks.

The remaining problems, however, were formidable: continuing high levels of sulphur dioxide, nitrous oxides and hydrocarbons and, of course, increased doses of carbon dioxide. There were also more discoveries of water being contaminated by nitrogen fertilizers, and still other forms of pesticide. There was continued dumping at sea of radioactive wastes, and much flood damage.

Mexico City, by 2000, may not only be the world's largest city but its most polluted. Although unemployment will be severe, some six million will be able to afford pollution-spewing cars. The city's air pollution will be aggravated by its high altitude and its location within a mountain range.

Increasing use of cars will exacerbate the air contamination problem in years to come, and they will contribute inexorably to the warming effect. Los Angeles is a victim of its endless sunshine, and so, for that matter, are many Mediterranean and Middle Eastern cities. Photochemical reactions with exhaust fumes create persistent unhealthy smogs; and in LA the smogs only clear for 156 days a year in the winter and autumn seasons.

Studies in West Germany showed in 1980 that city streets are still noxious places in which to spend any length of time. Trace elements discovered were vanadium, mercury and cadmium. Others, like iron, act as catalysts for other chemical reactions. Lead and bromine issue in ever-increasing amounts from car exhausts. Electric utilities, petrol refining and consumption, the mechanisation of agriculture, the increasing numbers of cars and buses, all add deleterious and often harmful gases to the air we breathe.

Invisible pollutants now rise to phenomenal heights, and travel great distances. The extraordinary longevity of the more modern pollutants arises from the way aerosols become disseminated throughout the atmosphere and even well beyond. Aerosols are airborne particles, and range in size from the molecular to visible specks of dust. They are now capable, however, of destroying vital life-sustaining gases that exist high in the stratosphere. It is the very tiny particles that pose the greatest threat to the integrity of the skies, because they remain airborne longer, and travel further, even as far as the North Pole. Many scientists hence conclude that while city air pollution is less in magnitude than it was in the 1950s, it is vastly more pervasive and damaging in ways not previously imagined.

One of the redeeming features of the heavier types of particle is that they pollute the land surface rather than the atmosphere, as they merely fall back to Earth under gravity. Particles also tend to stick together, and this aids the fall-out process. But this type of particle is fast disappearing. Many enterprises now regularly use smoke-removing equipment, or have changed to cheaper fuels. A

Weather changes (in percent) resulting from major urbanization in the Northern Hemisphere

Weather phenomenon	Annual	Cold Season	Warm Season
Contaminant volume	+1000	+2000	+500
Solar radiation	−22	−34	−20
Temperature (C)	+2	+3	+1
Relative humidity	−6	−2	−8
Wind speed	−25	−20	−30
Cloudiness frequency	+8	+5	+10
Rainfall	+14	+13	+15
Storm frequency	+15	+5	+30

report by the US Dept of Environment shows that the concentration of smoke in urban areas, in 1980, was down by 80% on 1960 levels.

On the other hand many lighter, atomized, aerosols are capable of creating hazy skies, and even smogs. The milkiness of some skies is evidence of the tendencies of some aerosols to scatter and diffuse the Sun's radiation. Lighter particles, wafting high over a dark vegetated surface, would reflect back more light, thus causing a cooling. But other aerosols do precisely the opposite. If they are darker than the underlying white ice sheets and snow banks of Earth they absorb more light and heat, and so warm up the atmosphere.

One of the disturbing characteristics of modern pollution problems is the growing predominance of carbon-based particles (see Chapter 5). Computer climate models suggest that carbon elements in the Arctic could affect the weather in the northern hemisphere, and could raise temperatures enough to distort the world's climate. Scientists were surprised to find that the grey haze in arctic regions, first noticed by pilots in the 1950s, was the result of a massive pall of pollution that had wafted up from the coal-based heavy industries in Russia and Europe. The pall was much heavier and denser than expected. Indeed, America's National Ocean and Atmosphere Authority, in a report, said that hydrocarbon pollution in the Arctic regions was 'on a scale and with an intensity that could never have been imagined, even by the most pessimistic observer'.

For these reasons a majority of scientists now believe that atmospheric aerosols of the warming kind are now more prevalent than the larger cooling dust particles. Stephen Schneider, one of America's top climatologists, believes that if concentrations of the latter kind predominate, then 'demonstrable climate changes could occur by the end of this century, if not sooner'.

The Shrinking Ozone Layer
A curious phenomenon occurs in the near space of Earth. Ozone is a form of oxygen with three oxygen atoms (O_3). But it is produced from ultraviolet radiation when ordinary oxgyen molecules (O_2) are split apart, to then re-combine into O_3. The ozone layer extending some 15 to 25 miles above the surface of the Earth is destroying and reconstituting about one-quarter of itself

continually. Yet enough of it remains to enable it to absorb the high levels of ultraviolet radiation from the Sun, and thus prevent us all from suffering from skin cancers or from other mutations, and from affecting biological life in ways that we cannot yet foresee. In effect, in an impressive self-correcting process that Nature is so fond of displaying, solar radiation first creates ozone, and the ozone in turn protects us from the worst excesses of that radiation.

Extra ultraviolet radiation would not, as has been suggested by some commentators, itself add to the global warming. But a shrunken ozone layer will allow more overall solar energy, radiating on a variety of spectra, to reach the planet's surface. And as the ozone layer will have a varying effect depending on its latitude, it can destabilize the weather in quite unforeseen ways.

As we will see later, ozone is both sinned against and a sinner itself, as it can damage vegetation. Hence we ought, at this stage, to highlight the importance of trees and vegetation in regard to human wellbeing. As any proverbial schoolboy knows, human beings breathe in oxygen and breathe out carbon dioxide. And trees do the opposite. He is thus well aware that if the Earth didn't have any trees or vegetation we would all (after a few thousand years or so) suffocate. Fortunately for us the world has not yet run out of trees, although if the reader persists to the following chapter he may well believe that this state of affairs will end within his own lifetime.

Nitrogen is present in about 78% of the air we breathe, and is even more vital to the human race. Much of the body consists of H_2O — water — but the molecules in the more solid bits of muscle, nucleic acids and enzymes all contain nitrogen. Nitrogen blends easily with oxygen to form nitric oxide (NO), and this is done continuously around the urbanized world inside vehicle engines. With an additional nitrogen atom the gas becomes nitrous oxide (N_2O), or, instead, with an extra oxygen atom, it becomes nitrogen dioxide (NO_2), and, with the addition of yet another oxygen atom, it becomes nitrate (NO_3).

However, like a lot of organic substances, nitrogen compounds are dangerous in large quantities. What is important is that nitrogen-based gases are able to destroy ozone. All this is quite natural and normal. But man is adding to nitrogen compounds by his use of fertilizer. And as fertilizer production and use is bound to increase as more food is grown (as the population expands), yet more nitrous oxide — the compound with the extra nitrogen particle, and hence the most damaging — will be pumped into the air, and more will get into the stratosphere. Within just 15 years, from 1954 to 1969, the use of nitrogen fertilizers in the US increased by 1,400%.

We are interfering with the ozone layer in other ways, too. Supersonic aircraft have been suspected for some years of ejecting oxides of nitrogen and sulphur dioxides, as well as water vapour, into the air. Over a period of years aircraft could double the existing amount of water vapour in the stratosphere, leading to a small rise in temperature, perhaps, according to some reports, up to 0.6C.

All this gaseous activity is yet further aggravated by Earth's worsening heat balance. We know that air temperatures decrease with height, and that terrestrial heat would normally be dissipated into the atmosphere. But in a hot city temperature inversions can occur where a trapped pocket of air can maintain its temperature while it ascends. If there is little surface wind an unpleasant smog hangs over the city.

A rather different situation ocurs when the air that carries pollutants is far warmer than the surrounding pockets of air, so they are carried higher. They then start to cross national boundaries, and other cities suffer as well. Eventually not only vast areas of the upper atmosphere are affected, but the Earth's climate also.

One reason why the top of the stratosphere is warm is because ozone is absorbing much of the Sun's radiation. At this height a heat inversion takes place. As the hot air rises the cooler pollution-bearing pockets of air become permanently trapped in the stratospheric zone. Nor can they rise further, so with high wind speeds the pollution is pushed around the world.

Fossil fuel combustion also converts nitrogen into nitrogen oxides, which in turn become airborne nitrates. Nitrates ultimately fall back to Earth and its oceans. The tropical Atlantic has the highest concentration of nitrogenous materials, as water bacteria are able to transpire using nitrates instead of oxygen. Some scientists believe that the oceans hold the key to the Earth's constant supply of nitrogen, either because of changes in the

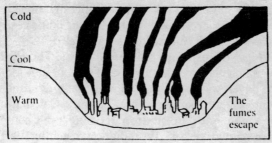
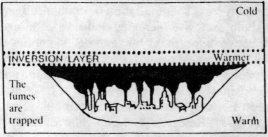

The heat-island effect is also made worse by photochemical smog. Rising warm air is sometimes trapped by a natural warm layer of air, known as an inversion layer.

Source: Philip Neal, Acid Rain, Dryad Press, 1985

biological activity of the oceans (and the way denitrification occurs), or through changing surface temperatures.

However, the most important danger seems to come from 'aerosol' spraycans. About half the cans currently in use employ compounds called fluorocarbons to propel the active ingredient (deodorant, hairspray, paint, etc) into the air. The gaseous propellant in spraycans was first discovered in the 1930s, after researchers found groups of chlorofluorocarbons (CFCs) that were cheap and chemically stable. They found they could be used in fire extinguishers, and were marketed as Freon-11 and Freon-12. Then, in 1950, they were first used in hairsprays.

In the meantime some atmospheric chemists discovered that certain forms of chlorine in the upper atmosphere were destroying ozone. A US government study said that chlorine could set off a chain reaction wherein just *one* chlorine atom could destroy 10,000 ozone molecules. What was puzzling was that there was no natural form of chlorine in the stratosphere, so it must have been man-made. Fluorocarbons, it later turned out, could easily be split up by ultraviolet light into chlorine particles.

So the finger of suspicion pointed to CFCs. The US National Academy of Sciences recently reckoned that a 16.5% reduction in the ozone layer was possible if Freon gases continued to be released into the atmosphere at present rates. Calculating on the basis that a mere one percent drop in ozone levels could cause a three percent increase in UV radiation reaching Earth, some rather startling conclusions were arrived at. Not only would the more harmful types of solar radiation be dramatically stepped up. Strangely, both excessive increases *or* decreases in ozone levels can deleteriously affect plant life, and cell structures can suffer degenerative changes. And any destruction or stunting of plant life would increase the amount of carbon dioxide in the atmosphere, which itself could react nefariously with CFCs to trap Earth's surface heat. Even by themselves, according to scientists from NASA's Langley Research Centre, chlorocarbons are very effective absorbers of infrared heat.

The problem with the ozone layer, first brought to the attention of the public by the mass media in 1974, shows no sign of abating. Indeed, it suddenly seemed to get worse in 1979 when satellite observations revealed a 'hole' above the frozen interior of Antarctica. In September 1983 some revealing truths about the ozone layer were published in *Nature* magazine by Dr. I. Brasseur of the Belgian Space Aeronomy Institute. A scientific satellite, the Solar Mesosphere Explorer, showed that the mechanisms which continually split up and reform ozone vary at different height levels. Below 60 kilometres the satellite showed a distinct seasonal variation in ozone concentrations. In one layer the density of the gas increased proportionately as the temperature climbed. But at higher levels the reverse was the case. At over 85 kilometres there was in fact quite a dramatic variation in the density of the ozone layer, even on a daily basis.

The satellite also discovered that an energetic proton burst from the Sun could sharply decrease ozone concentrations by changing the concentrations of ionized hydrogen, oxygen and nitrogen in the middle atmosphere. Much of the ozone, in fact, is right now being destroyed by solar bursts. Indeed, on the morning of one solar eruption the satellite observed that 70% of the ozone at a height of 78 kilometres was suddenly depleted.

By the summer of 1986 another development in the ozone saga occurred. American scientists were reported to be baffled by the

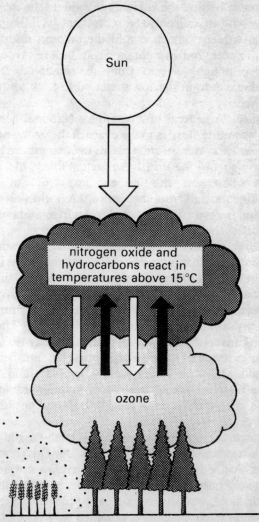

Compounds of nitrogen in the atmosphere can destroy ozone, and ozone itself can destroy trees.

Source: Acid Rain, John McCormick, Franklyn Watts, 1985

sharp drop in ozone levels, as the 'hole' had spread into the Drake Passage separating Antarctica from South America. The hole had by now become virtually the size of the US.

No scientist was prepared to make a snap judgement as to the prime mover behind the hole's appearance. It could have been a natural event, as it displayed a prominent cyclical characteristic, having a tendency to appear in the autumn months. For this reason the pollution hypothesis began to lose favour with some experts, although a team from the National Oceanic and Atmospheric Administration preferred to stick to the 'chemical input' theory.

The director general of Argentina's National Meteorological Service was quite clear as to the cause of the ozone hole, saying it was due to 'leaky refrigeration plants in the northern hemisphere'. And a member of the British Antarctic Survey, Mr. Joe Farman, put the blame on chlorine escaping from air-conditioned American cars, and from the rest of the world's air-conditioned offices and factories. He suggested that refrigerators be made more leak-proof.

Whatever the cause the enlarging hole was worrying since atmospheric scientists were uncertain as to the global implications. There was no shortage of scientists, however, supplied with expensive back-up equipment, prepared to investigate this alarming new development. The NOAA team flew into McMurdo Station in Antarctica, and a team from Wyoming University and the State University of New York launched thirty 19,000 cubic-foot balloons carrying ozone-measuring instruments. Scientists from the Goddard Space Flight Centre in Maryland launched a Nimbus-7 satellite to orbit both poles, and found that most of the decrease occurs during the long twilight as the Sun rises over Antarctica.

It later turned out that Antarctica was not the only spot where ozone levels were low. According to Donald Heath, a NASA scientist, Switzerland has suffered an average ozone loss of 3%, mostly in the past 10 years. Another hole, one-third the size of the Antarctic hole, was discovered over Spitsbergen, Norway, 700 miles from the North Pole. It extended over northern Europe to Leningrad.

Heath estimated an annual decline of nearly 2% in the northern hemisphere, with the largest losses occurring in February and

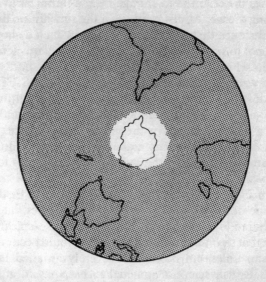

This satellite image of the ozone 'hole' taken in 1986 shows (very pale area, centre of picture) that the hole is now larger than Antarctica. The full implication of this phenomena has yet to be determined by scientists.

Source: Time magazine, 3/11/86

October. These observed changes, said Heath, were 'quite different from what has happened in the past'.

In the meantime new explanations about the role of the Sun in ozone creation and depletion were appearing in the scientific press. EEC scientists were reported as believing that the fluorocarbon explanation was becoming unfashionable, and instead tended to blame solar cycle variations, volcanic dust and even the jet streams snaking some 40,000 feet in the upper atmosphere. Scientists from America's Massachusetts Institute of Technology were also reported as saying that declines in ozone levels might have something to do with stratospheric 'upwellings'.

The scenario is a bleak one, because even if the ozone-destroying gases were halted overnight it could take 50 years to repair the damage, according to Mr. Farman. One major difficulty is that there are no natural mechanisms in the lower atmosphere to destroy chlorofluorocarbons. Scientists are still uncertain as to what causes the chlorine to change from a harmless to an active form during the winter, just waiting for the sunshine to turn it into a toxic substance that can destroy ozone in such a short time.

This is not to say that the authorities were not quick to react to the dangers when they became apparent. In 1974, shortly after the problems were first publicly debated, CFC11 and CFC12, the major industrial equivalents of Freon, were banned from 'non-essential' aerosol use in America and Scandinavia. The EEC imposed a 30% reduction in their use. Other sources of CFC, however, have continued to make up the leeway. Indeed the concentration of CFCs in the atmosphere continues to grow at 7.5% per year.

Early in 1986, a UN study, supported by the US Environment Programme and many other international bodies, had already been delivered to leaders of world governments. A clear warning was given that significant climatic changes would occur within the next 80 years unless emissions were severely curtailed. Dr. Robert Watson, a British scientist attached to NASA said at the time, 'Policy makers must be aware that if there is a change in atmospheric ozone, or in the climate, recovery will take centuries'.

In the meantime DuPont, the company that invented Freon, suggested that worldwide production be curtailed, and is presently, in coalition with other companies, searching for substitutes. The EEC has overcome its own doubts and has set a ceiling on production that leaves room for a rise of about 35% on present production levels.

Carbon dioxide, however, is the biggest rogue gas of all, deserving an entire chapter to itself. With or without other elements, it is the most important warming factor currently known to science. In the next chapter we shall be looking firstly at the human activities that are at this very moment pumping disastrous amounts of carbon dioxide into the atmosphere to dramatically raise the world temperatures within less than 100 years.

Chapter Four:

THE BALDING EARTH

The most serious environmental problem now facing the Earth is *deforestation*. Its effects, if left unchecked, are almost certain not only to bring about permanent ecological harm to the biosphere, but long-term — and irreversible — climatic change.

The situation in the late-eighties already looks ominous. The world's entire forestry and jungle reserves now cover only a quarter of the land's surface, with the highest ratio given over to South America, followed in order by Europe, North America, then Asia, and finally Africa. But up to half the world's original woodlands have vanished since 1950. Within the last 200 years Latin America has lost around 37% of its original tropical moist forests, south-east Asia about 38% and Africa almost 52%.

It is now estimated that a forest the size of Cuba is being destroyed each year. Brazil's forests have already been reduced by a quarter or more. The Food and Agricultural Organization of the United Nations (FAO) reports that more than eight million hectares of forests are annually eliminated in Asia. Sizeable parts of the world have now taken on the characteristically arid look of the Middle East. Even Britain has only 10% of its original primeval woodland left. And in just the last ten years Thailand has lost a quarter of its forest; the Philippines one-seventh in the last five. At this rate much of the Far East will have no lowland rain forest left at all by the year 2000.

The size of the problem can only be understood from an historical perspective. Some 5,000 years ago the world would have been unrecognizably densely wooded, with most human activities seemingly conducted in large clearings set within interminable forestland. Amazingly the Sahara Desert was once forested and inhabited. The glories of ancient Mali and Ghana in West Africa were legends in Medieval Europe. The land of Canaan was fertile and full of 'vineyards and olive-yards, and fruit trees in abundance' (Nehemia 9:25). So familiar are we with the desert regions of the Levant that we find it difficult to believe the Bible when it mentions 'the forest in Arabia' (Isaiah 21:13).

Great civilizations need immense fertile regions in which to flourish. The valleys of the Tigris and Euphrates supported thriving communities who became the virtual progenitors of western civilization. A massive irrigation complex fed rich lands which were the granary of the great Babylonian Empire. Pliny, the Roman naturalist and writer, tells of harvesting two annual crops of grain on this land. Today, however, less than 20% of the land of modern Iraq — old Mesopotamia — is cultivated. Ancient irrigation works now silted up, and rocky mounds, hint at the existence of long forgotten towns. Much of Iran is in a similar condition.

These civilizations destroyed, through profligacy, first their environment, and then ultimately themselves. The Sumarians probably prided themselves on being master tree-fellers. But their activities resulted in so much silt entering the Tigris and Euphrates, and pouring down to the Persian Gulf, that the two rivers were pushed 130 miles further south. In Mesopotamia, the cradle of civilization, tree felling caused massive soil erosion. Plato, speaking of Attica in Greece, said that the mountains were covered with trees. Roofs made from their timbers are still in existence.

Reference to Roman writers clearly shows that certain rivers such as the Danube and the Seine flowed much more sedately than they do now. And there can be little doubt about the reasons: the later removal of forests that used to regulate the run-off, and then the erosion of topsoil.

Around the world the story can be repeated a hundred times. Deforestation and its aftermath caused the downfall of the Sinhalese civilization on Sri Lanka. The Inca empire in Peru was forced to move to ever higher valleys after the felling of low-elevation forests and the salinization of agricultural land. It was this later difficulty in maintaining food supplies from an over-exploited soil that soon predicated the demise of the Inca civilization. The famed cedars of Lebanon have been decimated over 5,000 years of timbering. At one time the Greeks employed as many as 8,000 men, and 1,000 pairs of draft animals engaged in exporting the timber back to Greece, whose own forests had been largely cleared. The remaining cedars lasted until World War I, when the last few acres were decimated by Turks.

The absence of forests along the Mediterranean coast from Spain to Palestine is, too, largely the result of Roman exploitation. Typical of the region is the North Dalmation coast where the hills were once magnificently clothed with primeval woodland. The Romans, the Illyrians, the first Slavs, wreaked havoc upon the arboreal environment for decades. Indeed, until the dawn of the 20th century the razing of Europe's forests had never been conducted with such rapacity. By about 900 AD central Europe north of the Alps consisted of at least 80% forest, and by 1900 AD this had been reduced to 24%.

It was the areas of greatest population density that naturally made the most claims on the environment. Before the Industrial Revolution the chief building material for an advancing and complex civilization was timber. In Europe intensive agricultural practices coupled with heavy animal manuring made the cleared land even more fertile for a while than the forest soils it replaced.

The speed of deforestation matched the progress, or otherwise, of civilization. When the Roman empire disintegrated after the fifth century AD, forests reclaimed the abandoned fields. During the Dark Ages in Europe, and again following the horrific predations of plague and war in medieval times, forests started to come back. But not entirely, because by then farms had been erected in much of the deforested areas, and the growth in food production was to later fuel an industrial civilization. So much so that after 1500 AD timber shortages in Europe were becoming acute, and once again the forests were plundered. As the arboreal regions shrunk remorselessly wood became more valuable, and the ruling classes even attempted to preserve the woodlands by serving prohibitions on the wood-cutting activities of the peasantry. In Scotland the forests were cut down to serve as fuel for ironworks. The English needed much timber for their war galleons, and New Zealand lost 15 million acres of woodlands shortly after English migrants arrived there.

The Pilgrim Fathers and their descendents soon turned America's early forest empire into a memory. In the late 19th century, because of the dominance of wood as building material, as much as 85% of the woods of New England were being chopped down. By 1865 streams muddy with silt were commonplace, and floods were increasing. The mighty Mississippi soon became known as the 'Big Muddy'.

Washington and Jefferson were alarmed at what they saw taking place around them, and crusaded against destructive farming practices in word and deed, but to little avail. There was, after all, in a land as overpoweringly vast as America was to the early settlers, no real incentive. When one tract of land wore out, new land was always available just a little to the West. New technology, then as today, vastly exacerbated earth-ravaging practices. Yet more and more acres of cultivable land fell victim to McCormick's invention of the reaper in 1831, and the other refinements that followed. The self-scouring steel plow, for instance, from 1837 onwards, accelerated the westward march of agriculture.

China, with its massive population pressures, has suffered even more than the rest from wanton tree felling. At the time of the Shang Dynasty China was covered with trees. They were cut down ruthlessly as the need for more agricultural land became pressing. In northern areas trees disappeared to fuel kilns of the pottery and porcelain makers. The eroded loess, washed down as silt, made the lower plains fertile. But it also eroded that soil from elsewhere — transporting some 2,500 tons of the stuff annually. Soon it raised river beds and increased the frequency of flooding, both from the rivers and from the storm-waters rushing down denuded Chinese hillsides.

Slashing, Burning and Nibbling

However, Man's continuing intervention in the ecological cycles has now greatly speeded up the process of earth-balding and soil eroding. The very skin of the Earth itself is under attack. Climatic change, we now know, dried out the fertile Biblical lands of old. But a decrease in rainfall levels would certainly not have accounted for the rapid rise in aridity in the now inaptly named 'Fertile Crescent' if it were not for the over-exploitation of soil fertility by nomads, peasants and farmers.

There can be no doubt about the pernicious effect of humanity on the world's plant life. The ravaging of the Earth's surface has been going on for some thousands of years as Man burnt trees and undergrowth for fire, light and cooking, to improve mobility when hunting, and to increase grassland for pasturage. Tierra del Fuego, the 'land of fires', was named by Magellan in 1520 because of the extensive land burning near the tip of South America. In the Wild West the Red Indians hunted buffalo by driving them out of burning forests.

In many parts of medieval Europe over-grazing by sheep and goats has taken its obvious toll. Goats are highly destructive — they can insidiously reduce forests by nibbling away at the bark and sprouts of trees. Goats have often been preferred by the poor because they can survive on the sparsest of diets — even prickly pear cactus and poisonous locoweed. They are hardy, fecund, provide plenty of meat and milk, and can browse sure-footedly on mountainous steppes (some can even climb trees). However they are certainly a critical factor in reducing parts of Africa and south-west Asia to semideserts, and in turning 15th and 16th century Spanish agricultural land to the state it is now.

John Rich was one of the first geologists to spell out what the world's growing livestock population was doing to the global biomass. Writing as early as 1911 he said that New Mexico's abundant sheep (some 8 million in 1880) had plundered the vegetative cover and pulverized the topsoil of the earth. He recognized how easily animals could compound the follies of Man.

Let us take the case of Italy. The floods of the Arno have been a regular occurrence since the 14th century when the woodlands around Florence were converted to pastureland. Then overgrazing by farm animals ultimately turned the ground into baked clay. The earliest Arno flood was in 1333 when 300 were drowned. After that re-afforestation projects were urged, but ignored. Since then there has been a major flood every 100 years or so, and no re-afforestation is even now taking place. Indeed no serious attempt has been made to dredge or deepen the existing riverbed, and six feet of silt still remain from the devastating 1966 flood.

The last world war acted as a catalyst for Earth ravaging activities. Goats, firewood scavengers and the British troops in North Africa and the Far East put an end to the few remaining woods still standing, and left an exposed topsoil that soon became pitifully thin.

After the war, the Japanese were urged by the Americans to

improve their agricultural output by converting forests to farmland. The Japanese, having learnt from the follies of the Meiji period when deforestation had caused disastrous floods, had already imposed strict forest conservation laws. But in 1945 they were virtually obliged to fell, with the same results: floods and erosion. Before long the Japanese re-enacted their forest protection code.

As the soil is remorselessly eroded year by year by man and his animals so the world's peasant farmers advance deeper and deeper into the forest, slashing and burning anew, and the cycle is repeated. In Haiti slash-and-burn farmers have succeeded in converting most of the upland areas into rocky plateaux. Across sub-Saharan Africa millions of domestic fires have chewed barren gaps in the jungle. The Sahara Desert, in the meantime, expands by 100,000 hectares every year, made worse in recent times by severe drought conditions. As the trees become literally thin on the ground African women become more and more enervated as their working week is spent hunting further afield for firewood.

But this constant ravaging means that deforested areas are not allowed to recover naturally. The increasing shortage of firewood — a principal fuel for three-quarters of humanity — forces the exhausted women to hunt for crop wastes and dried-up excrement — anything that will burn — for cooking and heating. This, however, only denies whatever good soil is remaining the necessary nutrients for continued survival. The land on the fringes of the Sahara is so heavily over-grazed and plundered for its firewood that in the Sudan, for example, the desert has advanced 60 metres in just 15 years.

Again the pressures of population are at the root of the evil of deforestation. Worldwide demand for wood products doubled in the 1940–75 period. In 1945 US Forest Service lands produced 2.4 billion board-feet, but by 1970 10.8 billion board-feet were harvested. This is not to say that some commendable reversals in the trend have not taken place in America. Despite increased wood use and diminishing acreage, commercial forest growth rates in America were 14% higher in 1972 than in 1960. This has occurred, however, largely because of greater protection against disease, fire and insects. Tree planting has been carried out only spasmodically, as in Massachusetts. On the whole it has been grossly deficient, and done mostly with the faster-growing softwoods which can disturb the ground's nutrient balance.

But the worst postwar victim, by far, is South America. The 20th-century demand for timber is now largely to feed the newsprint mills. The US National Forest Reserve believes demand for timber and newsprint will double again by the year 2010. Prices of pulpwood in the meantime have been rocketing. Two-thirds of Latin America's forests have gone, or have been gravely diminished. Of Amazonia's six million kilometres of forest, as much as 10,000 km are cleared each year. Venezuela has lost one-third of its northern rain-forests during the period from 1950–75. At the present rate of denudation all Amazonia's trees could disappear in another 70 years.

Beef raising has grown apace, spurred on by rising prices in the developed world because of what is known as the 'hamburger connection'. As huge cattle ranches require more pastoral land, so again the forests take second place. In one case a multinational corporation was said to have burnt down a million acres of forest in the Amazon Basin to make way for a cattle ranch. The fire was said to be so big it was reported by a weather satellite as an impending volcanic eruption.

Other areas of the globe suffer from human rapacity. The West has an insatiable appetite for veneers and rare hardwoods — to make pianos, mahogany coffee tables and teak floors. The search for these marketable hardwoods is concentrated in south-east Asia. A modern chain-saw can slice through a tree in a matter of minutes. A tree-crusher can lop down trees and caterpillar tractors can cable-winch giant logs with ease. Indeed a whole hectare of forest, with the aid of modern technology, can be denuded of timber in about four hours.

Tree felling has been so successfully pursued that virtually all the lowland forests in Malaysia and the Philippines will surely be depleted by 2000. Thailand is already forced into the position of having to *import* wood products. Virtually all of Indonesia's lowland forests seem certain to have been exploited for timber by the year 2020, and this probably applies to Peru and Colombia. Logging concessions for Japanese multinationals (Japan is a major consumer of hardwood) have dried up. Japan, seeing the light, had to appeal to UNCTAD for a new deal for producers and consumers.

Acid Rain

The situation in Europe, however, has different antecedents. Thousands of square miles of forest in northern and central Europe now resemble a First World War battlefield, and a vast number of lakes have become devoid of life. But it is not the axe or the bulldozer that is to blame. Acid rain is said to be the cause, and has now become a burning political issue in Europe.

The acid rain problem is a product of atmospheric overload discussed in the last chapter. The prime mover is fossil fuel waste pumped from high smokestacks, ironically designed in the 1960s and 1970s to 'protect the environment'. Acid rain is a depressing reminder of how intolerable the pollution problem has now become. Like a virus, newer forms of contaminant constantly emerge just when particularly serious strains have been conquered.

Any schoolboy in a laboratory can reproduce the chemistry that causes acid rain. When coal or oil (or petrol) is burned, oxide compounds of sulphur and nitrogen are produced. These then change in the presence of sunlight; i.e. photo-oxidation occurs. This latter factor explains the seasonal variations in acid rain deposition. On a clear summer's day the conversion of sulphur dioxide (SO^2) and sulphuric acid is up to five times greater than on a winter's day. The insidious end-result of this oxidation process is the sulphuric and nitric acids that leech into the soil. It is known as 'acidification', and produces further chemical reactions on that part of the biota it falls on. Not just lakes and trees suffer, but rocks, buildings, industrial plant and even railways are all seriously corroded. The Germans are expecting a catastrophe within the next year or so as structures weakened by acid rain begin to collapse.

The acid does indeed fall like rain — it looks the same, feels the same, but the taste is quite different. Of course all rain — because of the presence of carbon dioxide in the atmosphere — is slightly acidic. But the additional ingredients in urban rain are testified to by many a hapless pedestrian who has been caught in a rainstorm by the smarting of his eyes. In small doses acid rain can be beneficial, helping to dissolve minerals and providing a lively tonic for jaded plant life. But, like most tonics, it can be extremely harmful when taken to excess.

And acid rain is now coming down to Earth in destructive quantities. The world's industrial superstructure now out-generates

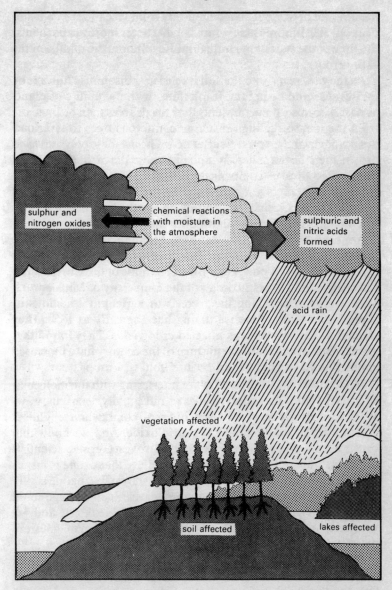

Oxides of nitrogen and sulphur can combine with water vapour in clouds to form nitric acids. It then falls as acid rain.

Source: John McCormick, Acid Rain, Franklyn Watts, 1985

the amount of SO^2 outgassed naturally from volcanoes by a factor of about 100. In some places rain is 1,000 times more acidic than at the time of the start of the Industrial Revolution, the middle of the 18th century.

As the rain seeps into the soil it releases poisonous aluminium, manganese and cadmium. It interferes with the ability of plants and trees to absorb the nutrients they need. Trees then begin to die from the top, so the high reaching coniferous trees so typical of Scandinavia and Germany suffer first. The needles soon start to go brown, then the branches sag, and pus weeps down the bark. Then, as the tree tries to compensate for starvation it sprouts weird looking twigs. Finally it succumbs. In the meantime the water in lakes looks strangely clear and clean. But this is because most microscopic plant and plankton life has been killed.

Atmospheric nitrates also contribute to the acidity of rainwater. Scientists know that the nitrate component of 19th century air was much less than it is now. Peter Brimblecombe of the University of East Anglia and D.H. Stedman of the University of Michigan hit on the idea of checking the records of agricultural stations in Europe and North America dating back as early as 1853. They wanted to see how nitrates affected crop yields. They found that nitrate concentrations since the turn of the century have increased about 10-fold in the US and about 5-fold in Europe.

In fact, both man and nature are interfering with the molecular structure of the atmosphere in diverse, but equally nefarious, ways. We have already seen how the internal combustion engine is creating additional amounts of nitric oxide. And we know that ozone is a tri-atomic form of oxygen. Now West German scientists at the Institute of Ecology in Munich believe that ozone is just as much a tree killer as acid rain. Tests have shown that ozone is very damaging to the cells and structures of leaves. Such damage could allow the essential magnesium in the leaf to leak out and get washed away by rain. Eventually the tree suffers from magnesium deficiency and dies.

This probably explains why trees in the Northern Rhine area are dying en masse in spite of the soil being alkaline. Scientists, in fact, have found that the greatest tree damage occurred where the ozone levels were high, because car exhausts were making ozone come down to Earth in the form of photochemical fog.

The acid rain phenomenon is now whipping up global

resentment. Sulphur and nitrogen oxides are crossing international boundaries. Like the 'aerosol' problem discussed in Chapter Three the airborne oxides are forced to travel great distances on the prevailing winds. It is pollution by proxy. Most of America's acid disposition comes from Canada — as much as 500,000 tons of it annually. It was reckoned that a giant stack looming 1,250 feet above an Ontario nickel smelting plant contributes at least 1% of the world's sulphur dioxide emissions. Norway received an unwanted gift of 60,000 tons of this vitriol from Britain. Most of the remaining 190,000 annual tons also came from abroad. Svante Oden, a Swedish soil scientist, suggested that Sweden was being 'bombed' by acid rain producers in a 'chemical war'.

In the meantime the Swedish Forestry Commission completed an emergency survey of 10 countries, and said that in the worst areas some 50% of native trees were found to have suffered visible damage, with 15% almost denuded. Large numbers of pine trees now lay dying in Scandinavia. Mr. Mats Segnestam of the Swedish Society for the Conservation of Nature said, to a flurry of media publicity: 'The acidification of land and water is perhaps Europe's most serious environmental problem in the Eighties'. Dr. Lars Overrein, the head of the Norwegian Institute for Water Research, pointed out that in south Norway, an area the size of the Netherlands, freshwater lakes had lost more than half of their fish stocks between 1940 and 1980.

It is clear that the circulation of winds in the northern hemisphere is no respecter of nations. Britain, suspected as being the main offender, was nevertheless thought for years to get off scot free as her atmospheric effluent was lofted far and wide across north European skies. A recent study in Scotland, however, shows that rainfall is at least 15 times as acid as normal. This proved what had long been acknowledged by meteorologists; that the southeasterly winds of central Europe and the Ruhr were an even match for the westerly gales blowing sulphur dioxide from the British Isles.

In recent years Germany has suffered greatly from acid rain. German forests have for centuries inspired poetry and folklore. Sadly, some 30% of deciduous trees in central German forests are now doomed to become a folk memory. Already an area the size of Belgium is lost, and twice that much is damaged or moribund. It is estimated that as much as 80% of Germany's vulnerable fir trees are affected. Fresh reports of diseased German trees arrive monthly: 64,000 hectares dying in Baden-Wurtenburg (where the famous Black Forest grows) in the spring of 1982; 300,000 hectares found in Bavaria one year later, and so on. After Chancellor Kohl of West Germany said: 'The time for action is running out', some hurriedly improvised measures to oblige German power stations to cut back on SO^2 fumes from 3.6m tons a year to 1.6m by 1993 came into force. They were immediately condemned by the Green Party as 'pitifully inadequate'.

No European forest is too remote to be affected by acid rain. By the late 80s the problem showed no signs of abating. Indeed, it seems to be spreading. Parts of Czechoslovakia and Hungary are now in a serious condition. Farmers across the Continent have spent a small fortune on lime to try to combat the effects of the acid. The financial losses to the European community are massive. In Sweden alone farmers have spread 50,000 tonnes of lime on their land. The cost to Scottish landowners of lost production is something like £25 million. The Dutch need to spend £6.6 million yearly on repairs to buildings caused by acid rain. Annual repairs to Cologne Cathedral alone cost £1.5 million.

In America, according to the EPA, damage to buildings costs some £4,000 million a year. US ecologists are saying that fish in 100 lakes in the Adirondack wilderness of New York State have died. According to the environmentalist Erik Eckholm, in his UN-funded book *Down to Earth*, the average American raindrop is between three and 30 times more acidic than usual. In West Virginia a storm dumped rain some 4,000 times more acidic than it should be — sourer than lemon juice. President Carter in 1980 said that the after-effects of acid rain was one of the gravest environmental threats, and ordered a £50 million 10-year investment programme.

The cost of remedial measures, however, would be prohibitive. The Americans, already with the most stringent anti-pollution laws, reckon that slashing SO^2 emissions by 50% along the eastern seaboard might approach £5m a year. One UN investigation said it would cost £56 million simply to cut Europe's emissions by 50 to 60% a year.

In 1986 the British Government for the first time agreed that

Britain was responsible for some acid rain damage to Norway. The Government took note of what its own Dept of Energy was saying: that unless something was done soon SO^2 emissions might rise by up to 30% between now and the year 2,000. The CEGB has agreed to spend £600 million on a limited programme to clean up SO^2 emissions from three large power stations. But the Dept of the Environment is far from certain that this programme will noticeably reduce sulphur dioxide emissions up to the year 2000.

Chapter Five:

THE GREENHOUSE EFFECT

Man is creating a Hothouse Earth, and his chief method of bringing this about has been spelled out in the last chapter. The felling of trees and the destruction of vegetation on the scale outlined is as big a threat, if not bigger, to the world's climate than the Heat Island Effect, or pollution of the upper atmosphere. For when fossil fuels (in the form of dead and compressed vegetation) are burned, and when living trees are hacked down, massive amounts of carbon dioxide (one element of carbon mixed with two of oxygen — CO^2) are released into the atmosphere.

The human appetite seems insatiable, and our consumption of fossil fuels has been little short of prodigious. Since the time of the Industrial Revolution some 400,000 billion tons of CO^2 have been released into the air. And since the turn of this century we have been using so much coal, oil and petrol to fuel our economies that we have increased the amount of this particular gas by 20%. Nowadays an estimated 5½ billion tons of carbon molecules are pumped into the ecosystem every year. By evolutionary standards these are phenomenal amounts.

Many people are astonished to learn, for instance, that the coal now extracted from the earth in just one year is the end result of 400,000 years of natural deposition. One coal pit in America using a 500-ton electric shovel can scoop up 30 tons of coal in one fell swoop. The pit will eventually produce, before exhaustion, 350 million tons. This is over four times the amount produced in the whole of Britain during the year 1860.

The decimation of a forest which took 50 or 100 years to grow means that all the hydrocarbons in the wood are released into the ecosystem within literally hours. This is an unnatural and, ultimately, highly damaging phenomenon. Scientists are only now just beginning to fathom the consequences. We must remember that the carbon dioxide-oxygen balance in the atmosphere is controlled by plant life. It is estimated that ocean and land plants consume 500 billion tons of carbon dioxide every year, converting it and water into free oxygen and organic matter. If there are profound changes in the distribution of Earth's vegetation there will be commensurate changes in the abundance of free oxygen in relation to CO^2, a good percentage of which is also emitted by volcanoes and living creatures, including humans. Eventually a critical point will be reached. As some forest clearing is done to increase the acreage of arable land the balance is redressed somewhat as other CO^2-absorbing vegetation takes over. But we have seen that growing desertification and soil degradation means that there is still a continuous net loss in the world's stock of vegetation.

Our Carbon-based world

The carbon atom has enormous versatility, and is able to swap up to four of its outer electrons, and thus combine with a wide range of terrestrial atoms. One carbon atom can combine with four of hydrogen to become a hydrocarbon, and by piling on other carbon atoms can create large organic molecules. Hence carbon is the basis of organic chemistry and biological life itself.

However, some compounds of hydrocarbons can be volatile. Propane, for example (C_3H_8) and butane (C_4H_{10}) are part of the paraffin family. By swapping one of the hydrogen atoms for one of oxygen, the alcohol group of families can be formed. Alcohol compounds are similar to the paraffin and benzine families, and this explains why alcohol can be burned in some internal combustion engines.

It is easy to see how poisonous carbon compounds can become, and how they can alter the molecular balance of atmospheric gases. CFCs, of course, are one carbon-based compound. Just one carbon atom bonded to one of oxygen gives us carbon monoxide (CO). When CO is burned in air, it can produce carbon dioxide.

Carbon dioxide can also be produced quite naturally by organic living matter. Some of the 100,000 million tons of plant material produced annually by photosynthesis in the Earth's biomass is eaten by animals. The carbon then moves further along the cycle.

It is exhaled back into the atmosphere as carbon dioxide, and is used in turn by vegetation, which takes in the carbon molecules, plus hydrogen from water molecules, to make carbohydrates for regenerating their tissue. Plants in turn release the oxygen content back into the atmosphere, enabling animal life to breathe.

None of this, of course, could be done without the benign presence of the Sun. Through the process of photosynthesis the simple molecules are rearranged to form the chemical bonds between the atoms in the larger molecules.

But not all atmospheric carbon dioxide is released by the breaking down of carbon molecules by living matter. The *decay* of dead matter (and the burning of dead matter) performs a similar function. Here arises the danger with deforestation on the present scale. Micro-organisms of decay can break down organic substances into their simpler gaseous components which then compound with atmospheric oxygen — hence carbon dioxide from rotting trees, and even animal corpses.

Our planet is, in effect, a carbon recycling machine, and its main climatic and biological features are determined by this. We know that the tension between the gravitational pull of large bodies and centrifugal force keeps the planets and moons in their regular orbits. The same principle applies to Earth's atmosphere. What prevents the oxygen and nitrogen particles from falling to the surface is the convection heat in Earth's warm, equable climate. But Earth's temperatures have not always been constant, or even warm. Apart from volcanic eruptions and occasional periods of plasma-induced states from massive celestial collisions, it was the core of Earth that was molten, not the surface. During cooler solar periods atmospheric gases had a tendency to fall back to Earth.

Hence the early CO^2 that was in the atmosphere drifted back to the surface — more particularly its oceans — and ultimately fused with the ocean floor sediment. When the sea floor sank beneath the edges of colliding tectonic plates, the carbon became buried even further into Earth's interior. Ultimately the carbon molecules would get forced back to the surface again when volcanoes erupted explosively into the atmosphere. According to James Walker and Paul Hays at the University of Michigan it is this carbon recycling phenomena that accounted both for the unusually warm epoch during the Cretaceous period, and for the warmth of a primordial

The Greenhouse Effect

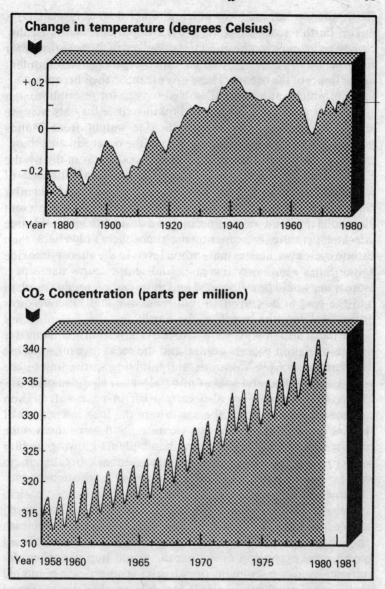

The upper diagram shows how temperature has been rising since 1880. The lower diagram shows the annual peaks and troughs of CO_2 emissions

Source: Science Digest, February, 1984

Earth under a Sun that was much dimmer than at present.

The Earth's seas, therefore, are an important feature in the carbon recycling hypothesis. CO^2 is dissolved in the sea to be used, with the aid of sunlight, by single-celled organisms living in the upper layers of the oceans. These tiny creatures then become food for fish, which can use the dissolved oxygen for respiration.

The seas are thus a friend of Mankind, since they are able to absorb the bulk of the carbon dioxide output from Man's profligate activities. However, the role of the oceans in absorbing carbon is the one still imperfectly understood factor in the whole Earth-warming debate. For example, if the Earth warmed a fraction the increase in evaporation from the seas would increase cloud cover. Oxygen in the upper atmosphere would absorb the ultraviolet rays, and water molecules and dust particles would also affect temperatures by scattering the atmosphere's blue light. The carbon cycle also dictates that carbon levels in the atmosphere fall as the Sun's luminosity increases and temperatures rise, since more water would be evaporated and more carbon would dissolve into the seas to be converted into limestone and ultimately be driven back into the bowels of the Earth.

The Earth also stores carbon in rock. As terrestrial temperatures rise, an oxidation process occurs, and the rocks give off some of their carbon as CO^2. Volcanoes and bubbling spring water (the kind that is bottled and sold as mineral water) also give off CO^2. Carbon-bearing ions are also carried off in river water. This carbonated water reaches the sea, where the ions interact with skeletal remains of tiny marine creatures, and turns them into calcium carbonate. On the other hand plants growing in the oceans would tend to have a negative feedback because they would consume more of the carbon, enabling more atmospheric CO^2 to be transferred to the oceans.

Then, as the warming continues, scientists have discovered further feedbacks. Wallace Broecker, a geochemist at Columbia University's Lamont Doherty Geological Observatory, believes that thermal expansion of the oceans would trap phosphorous nutrients in coastal sediment, thus depriving other ocean plants of food, some of which would die off. Hence less carbon dioxide from the air would be absorbed, and the cycle would be perpetuated. Other scientists believe that organic life plays an important role in the Ice Age cycles. Ocean life would tend to thrive in the warmer

The Greenhouse Effect

epochs, overpopulate the seas, and die as soon as the oxygen became depleted, and stop sucking CO^2 out of the atmosphere.

So why, exactly, does carbon dioxide warm the Earth? The simple answer is that it has the capacity to absorb most of the Earth's infra-red heat that would otherwise escape into space. When the molecules are hit by heat radiation, they turn the energy into motion, oscillating back and forth, vibrating energetically. This has the effect of re-radiating energy back to Earth. The analogy with a greenhouse is not quite aposite, since the high temperatures inside a greenhouse have more to do with absence of surface winds that would normally draw off a lot of the heat. But it is a useful analogy, easily understood and now almost common parlance, and reminds us of an important truth about the way heat balances are maintained or altered on rocky planets with atmospheres.

For the reality of what excessive amounts of CO^2 can do to celestial orbs like Planet Earth is grimly evident to every astrophysicist. Venus, with a temperature in excess of 480C at the surface, has a largely carbon dioxide atmosphere. True, Venus is closer to the Sun. Yet if it had the same levels of CO^2 as the Earth its temperature would plummet to around 180C. In many regions,

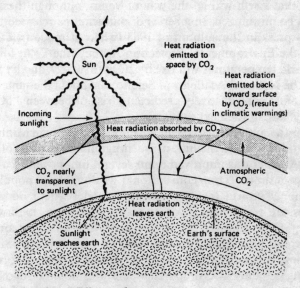

How the Greenhouse Effect works.

especially the polar regions, temperatures similar to those of the equatorial regions of Earth are achieved. Similarly if Mars were given Earth's atmosphere, its surface temperature, at present well below zero, would rise some 40C, making it bearable to Earth-like creatures.

Measuring the warming

The warming properties of carbon dioxide were discovered in the late 1800s by the Swedish scientist Svante Arrhenius. With a brilliant flash of insight he was the first to compare the process with that occurring in a greenhouse. Gradually the greenhouse phenomenon became established as scientific theory, soon to become a scientific fact. It was in 1957 that the greenhouse effect was first monitored systematically. Charles David Keeling, an oceanographer at the Scripps Institute in America, began the first contemporary carbon dioxide research programme from atop the Mouna Loa volcano in Hawaii, in the middle of the Pacific Ocean. Continuing to monitor CO^2 over the years, Keeling has shown that the gas in the atmosphere has risen from 315ppm in 1958 to over 345ppm today, a rise of nearly 8% in 27 years.

Will the warming trend continue? Scientists, mercifully, are not saying that Earth will go the way of Venus, although the threat cannot be minimized. In a series of publications released to the world's press in the autumn of 1983 (where it made front page news), the Environmental Protection Agency (the EPA) forecast that world temperatures will rise by 2C by 2030 AD, and by 5C by 2100. The National Academy of Sciences, at about the same time, rather more conservatively predicted a rise of between 1.5C and 4.5C by 'late in the next century'.

And in a more recent assessment, published by Irwin Mintzer of the World Resources Institute of Washington in the spring of 1987, the EPA's data was improved upon several times over. In a 'base case' scenario reflecting conventional wisdom about population and energy growth (i.e. a pessimistic model where no global policies are implemented, or at least acted upon, to slow the buildup of the greenhouse gases), the world's 'warming commitment' could actually increase by 8.6C on what it is now.

If these single digit figures don't sound much, it is worth recalling that only a 5C drop in temperatures would be enough to bring on an Ice Age comparable in severity with the last Ice Age, when the ice sheets reached as far south as Bristol. Furthermore the thawing of the Ice Age was much more dramatic in its effect, causing a catastrophic flooding around the world that probably formed the basis of the Great Flood legends of many ethnic cultures.

Average global temperatures have risen about one degree C in the past 90 years. During the first half of the century there were consistent and prolonged warm spells in the northern hemisphere. Some science writers, like John Gribbin, believe that there has been some return to more erratic and extreme temperatures since then, reflecting the norm of most of climatic history. But since about 1960 the upward trend has continued to the present day.

Indeed, the National Academy of Sciences believes that the global climate is already too warm for ecological comfort. Within just 20 years from now, they estimate, there will be a more noticeable rise in temperature. And yet in spite of the statistical forecasts few experts are sanguine enough to spell out exactly what will happen. Weather and climate are incredibly complex. For example, in a wide-ranging EEC project on the social and economic effects of a warming, the project writers admitted that they could not foresee whether rainfall in Europe would increase or not.

This might imply that most of the climate predictions could be taken with a large pinch of academic salt. The mass media inevitably wish to sensationalize the worst-case scenarios of global warming, and eliminate any qualifying clauses and concentrate on the purely speculative conclusions. But climate modeler James Hansen of the Goddard Institute for Space Studies, in New York City, exemplifies the attitude of most scientists in this field by confirming that the reports published in the press were 'within the range of plausibility'.

Computing a Greenhouse World

How is it possible to *know* that the greenhouse gases warm up the atmosphere? Might not the warming be due to astrophysical reasons?

All theories need to be tested empirically or experimentally. Ideally one would find a twin Earth in the solar system and simply pollute it, and see what happens. As this is impossible the scientists instead build a 'model' and pollute that. But even a scaled-down laboratory model would be inadequate, since no artefact can hang in space independently of Earth's gravity. So a mathematical model is devised instead, and the model is 'polluted' mathematically.

One of the success stories of recent decades has been the growing sophistication of microchip computers. A great deal of statistical information gleaned by science-based organizations like the EPA from studies of world fuel consumption is fed into the computers. The climate modelers have actually succeeded in developing huge electronic models which can be tilted so that geographical regions experience more or less solar radiation. Then the model will correspond to real astronomical conditions. After that the CO^2 input is added. Vital facts such as the temperature of the oceans and the skies — even information about air bubbles trapped in glacial ice — are all used.

What the computer is able to do most effectively is to juggle an astonishing number of variables. The drawback is that the computers, as a result, come up with a much wider range of predictions than would be preferred if the argument is to be drawn in bold, unequivocal, contours. Inevitably a great deal of speculation arises in terms of *how warm* the climate will become, and when, and how much warming will affect global weather patterns and the ice packs.

One of the most valuable sources of computer information is the Mouna Loa monitoring base on Hawaii. After a great deal of number crunching, conglomerations of equations were produced from which predictions about future weather trends are made. The problem is that the Earth orbits the Sun and spins on its axis. There are other complex variables in regard to winds, clouds and rainfall patterns. As clever as the computer is it can never, it seems, cope with the unfathomable complexity of the real Earth. As one climate modeler based at California's Lawrence Livermore Laboratory, and a contributor to the NAS report, admitted: 'A computer model as complex as the real world would probably be as large and as incomprehensible as the real world'.

One cause for optimism arises from the brilliant way in which computers can handle historical data. Many 'general circulation models', which include data on·solar radiations, atmospheric gases and the way temperature is redistributed across the globe, have successfully worked backwards. For instance, when CO_2 figures for the past one hundred years have been fed into the machines, the warming that has in fact occurred was predicted. This is proof that the computers can equally well extrapolate future trends from existing data.

But how do we know the CO_2 figures are correct? The key variables, to be converted into mathematical abstractions, are input by humans; but they may be either misunderstood or incomplete. For example, tracing the effect of CO_2 means that the computer has to be fed a detailed knowledge of the carbon cycle with all its byzantine complexity. It must understand its sources, its present and past compounds. It must know the rate of decomposition of plants and the nature of the gases emitted from volcanoes.

Let us, for example, take one recently discovered anomaly. Charles Keeling now believes that the increasing amounts of carbon dioxide are, paradoxically, having a beneficial impact on plant life. On the face of it the 'Greenhouse Effect' may be becoming just that. Keeling and his colleagues at the Scripps Oceanographic Institute at La Jolla, California, came to this conclusion after discovering a curious variation in CO_2 oscillations. Clearly, the level of the gas fluctuates according to the seasons. But the magnitude of the variation appears to be increasing too, and this is confirmed by other data from Alaska and from a North Pacific weather vessel.

Plants, of course, 'feed' on carbon dioxide, photosynthesizing carbohydrates to maintain their vigour. They thus become more active with rising CO_2 levels — absorbing more during the growing seasons and emitting more during the decaying months of autumn and winter. This phenomenon might, too, be confirmation that the biomass is decreasing through de-vegetation practices, thus releasing more CO_2, and in turn invigorating the remaining vegetation. But a part of the increase might, reckons Keeling, be due to the way in which the warming gases that overwhelmingly predominate in the northern hemisphere drift over to the southern hemisphere. This feature tends to increase the strength of the annual oscillation. The global warming is also playing its part in boosting plant growth.

The oceans remain a conundrum, perhaps because they are such a dominant geographical feature.

It is generally agreed that it would take literally a millenium for atmospheric heat and carbon dioxide to circulate from the uppermost to the very lowest parts of the ocean, and for the ocean water near the poles to drift down to the equator. The seas of the world in fact act as a giant thermostat, distributing solar heat across the globe, and account for the variations in the world's climate.

The British Isles, for example, are bathed in the warmth of the Gulf Stream to make the climate much milder than the latitude would suggest. Land-locked Russia, on the other hand, suffers from harsh winters, short summers and unpredictable rainfall. Charles Keeling's conclusion is that the oceans have a positive feedback characteristic that could cause a slowing of the greenhouse effect.

Another important natural variable is the atmosphere. Robert Schiffer, manager of NASA's climate research programme, believes the amount of cloud cover over the Earth is crucial to the equation. The warmer the surface, the more evaporation there is to form into clouds, and the more clouds there are the more Earth's heat is trapped. The alternative theory decrees that excessive cloud cover cools the Earth by blocking out sunlight.

New knowledge on the subject is accruing annually. On the first flight of Spacelab aboard the Columbia space shuttle of December 1983 a spectrometer measured water vapour in the tenuous stratosphere. Considerable research is at present under way, spearheaded by data gleaned from satellites that should help produce an atlas of global cloudiness and its impact on the greenhouse effect. Other investigations are concerned with changes in the Earth's orbit and the variability of solar radiation.

Then there is the problem of other greenhouse gases mentioned earlier. Nitrous oxides, methane, ozone and CFCs all counted now in parts per billion rather than parts per million. They all compound the menace of the carbon gases. According to Jerry Mahlman of Princeton's Geophysical Fluid Dynamics Laboratory, anthropogenic gases are pouring into the skies so fast 'that it is absolutely unknown what will happen'.

Then there is the role of vegetation. CO_2 records don't simply follow a linear path, but vary from one high point to another each year. Hence another complication is the existence of the natural cycle of the seasons. In autumn the falling leaves and decaying matter release more carbon dioxide. This is when the gas level is at its peak. Charles Keeling believes that these seasonal oscillations are getting wider.

The Warming — and Beyond
What, finally, would be the end-result of a greenhouse warming as prophesied by the experts? The EPA scientists hinted that we could get into trouble in ways barely imagined by earlier generations of Doomsday writers. It is too late, they maintained, to do much about it, even if a total ban on fossil fuels were imposed now. While other government studies have warned of the certainty that it *will* occur no matter what we do about it! Mr. John Hoffman, director of EPA's strategic studies staff, says that the warming is 'neither trivial nor a long-range problem'. A total ban on coal consumption around the world would perhaps delay the worst effects by some 15 years. Even a 300% tax on fossil fuels would only retard the warming for five years, at the expense of universal economic upheaval. A spokesman for the NAS said: 'Man-made emissions of greenhouse gases promise to impose a warming of unusual dimensions on a global climate that is already unusually warm'.

The EPA report, some 496 pages long, was the most specific and hence the most alarming (see Part II of this book). The polar ice caps and alpine glaciers could melt substantially, causing a two foot rise in sea level by the end of the century. Many coastal communities across the globe would soon be inundated. America's ability to maintain the role of the world's grain exporter would be seriously threatened as the drying out of fertile areas continued apace.

One blessing is that the much publicized drought-plagued regions of North Africa will get more rain, and could even become grain exporters. In the meantime, however, Dust Bowl conditions could return to America with a vengeance.

PART TWO

THE COMING FLOODWAVE

Chapter Six:

THE WARMING WARNING

The Deep South of America in 1986 suffered a gruelling heatwave that was superimposed upon the worst drought in the south's history. Throughout the summer thermometers rarely fell below the mid nineties, entirely drying out the land. Over thirty people died of the heat, and the governors of six of the southern states had to ask Washington to declare the region a disaster area. In the meantime hundreds of thousands of chickens suffocated in their sheds. Fish died in their shrinking ponds and rivers, and crops shrivelled on such a scale that their loss was estimated at $2 billion. Temperatures in North Carolina stayed at above 110F for weeks. Much of the state's staple crops of tobacco, corn and cotton was ruined.

But the summer of 1986 maintained the same extreme weather patterns that the world had been experiencing for the past decade or more. There were prolonged heatwaves in southern Europe. France had its fourth hot summer in a row. Temperatures in Turkey were in the upper nineties for literally months on end — at least 10F up on the average. There was a prolonged drought in Russia. Even Britain had above average temperatures with heatwaves at the beginning and end of the summer — thus giving the country three unusually warm summers out of four.

However, although recent summers seemed to be getting hotter, winters everywhere seemed to be getting much colder than usual. February in Europe was becoming a month to dread — invariably it was bitterly cold. In 1986 Britain experienced its second coldest February of the century. February in Europe in 1985 was the coldest in living memory. Most of the Baltic Sea froze for the first time since the war, making it possible for Poles to walk to the Swedish island of Bornholm and for Swedes to stroll over to

Denmark. Rumania's temperatures touched minus 40C, with most homes without electricity for three days a week. Temperatures in Paris were the lowest since 1894. In January 1985 the weather was so cold in sub-zero Washington that the presidential inauguration ceremony — for the first time in history — could not be held out of doors.

But the 1985 winter in the northern hemisphere was true to form. In the period 1983/4 a blizzard in Denver produced 45.7 inches of snow — 20 billion tons over a 6-day period — which crushed 14 buildings flat. A chain of winter tornadoes claimed 80 lives in the eastern zones. Both Britain and America had not experienced a normal mild spring for years, with hail in 1982 costing Texas cotton farmers $2.2 billion, and cold northerly winds in 1985 retarding the root vegetable growing season in Lincolnshire by a whole month. In mid January 1987 yet another record cold spell with sub-zero temperatures and blizzards afflicted Europe, taking an increasing toll of lives and disrupting transport systems and business activity.

But how can freezing temperatures be evidence of a global warming as predicted by the Greenhouse Effect? The answer is that the Effect strengthens high pressure atmospheric systems over the Northern Hemisphere, and allows the much colder easterly and polar winds to block off the warmer prevailing westerlies coming in off the Atlantic. Hence cities used to cooler regimes will not simply assume the temperature gradients of urban regions in more southerly zones just because thermal pollution is rising. For instance it was once suggested in a front-page article in the New York Times by John Hoffman of the EPA that New York would soon enjoy the balmy climate of Daytona Beach. Instead, climate warming models hint that New York would still be buffeted by the vicious winter storms plunging down from Canada.

Meteorologists refer to these stationary ridges of very high pressure as 'blocking highs'. More officially they are known as circumpolar anticyclones. Scientists are still not certain why they occur or what makes them suddenly break down. One theory suggests that northerners suffer from these blocking highs because the distribution of oceans, land masses and mountain chains is not the same as in the southern hemisphere. East-west sea surface temperatures in the North Atlantic also play a part. In a sense, although a blocking high effectively halts milder weather patterns in a kind of pincer movement, computer models of the climate suggest that the highs are in fact sustained by the outlying low pressure systems around both sides of the affected area. The whole phenomenon is virtually self-perpetuating, with additional energy being pumped into the anticyclone, and this helps to keep it going.

In the meantime revealing proof that the world's weather was becoming strangely erratic came in May 1984 when the Commercial Union insurance company of Britain hinted that it might have to raise its premiums if recent weather patterns were maintained into the future.

The company sustained massive losses of millions of pounds after damage around the world caused by storms, gales and floods were met. A spokesman for the company said that during the last decade much greater extremes of weather had been experienced, and the earlier pattern of a bad winter every five or six years was being reduced to every two or three years.

General Accident, another insurance company, said that their underwriting account for the UK showed an overall loss of £31 million, up from £19.3 million on 1982's figures. Claims for losses due to weather damage in January 1984 alone amounted to £18 million, pushing the group's overall loss of £16 million, compared with a comfortable profit of £8.3 million for the same quarter in 1983.

Thomas Karl, of the National Climatic Data Centre in Asheville, USA, believes that the stable postwar weather trend up to the late 1970s, and the later sudden breakdown of this trend, was 'striking'. With two other senior meteorologists from the American National Oceanic and Atmospheric Administration he published an alarming scientific paper confirming just how erratic the world's winter weather has been in the past few years. The past eight winters saw three which were much colder than usual (1976/77-1978/79), with three consecutive winters when it was warmer (1979/80-1981/2). Such a combination of six abnormal winters would not be expected to occur for more than 1,000 years.

It is clear, then, that the climate is displaying signs of acute instability. It is teetering from one mode to another with astonishing swiftness, and is baffling experts and laymen alike. One British journalist who wrote in April 1983 that the previous five years proved that Britain had been experiencing wetter than average weather was soon obliged to record that the country had been blessed with two hot dry summers in a row. Suddenly there was much talk about the Greenhouse Effect.

Yet the harsh winter cold snaps of 1978/9, 1981/2 and 1985 also, and without anyone noticing the contradiction, revived speculation about an imminent new Ice Age.

The Human Factor

And yet countless inhabitants of Africa might be forgiven for treating the idea of an Ice Age as a sick joke, so parched had their homelands become. As the winter of 1984 drew near in Europe, television screens suddenly became full of horrifying pictures of emaciated drought-afflicted refugees from Ethiopia. Relief experts were reporting despairingly that millions of acres of African agricultural land had been destroyed, and more than ten million children were likely to die.

Already some 30,000 had died in Mozambique and hundreds more in Angola, Botswana, Lesotho, Malawi and Zambia. As a series of reports from the United Nations Food & Agricultural Office made clear, some 150 million people, or more than one in three, were having to endure widespread malnutrition and death. This was in addition to the 125,000 who died in the 1968-73 drought, and which had affected some six countries in the Sahel region.

Even so, was Bradford Morse, a leading spokesman for UNFAO, right in declaring that the African famine was the 'worst natural disaster in history'? In an appropriately timed book, entitled *Natural Disasters — Acts of God, or Acts of Man?*, Anders Wijkman and Lloyd Timberlake conclude that disasters are increasingly becoming man-made phenomena, due largely to environmental resource mismanagement.

From the pages of this slim but meticulously researched study, prepared with the aid of many international disaster relief agencies, it is now clear that overfarming has played a key role in sustaining this horrendous African drought. Indeed the link between high population growth and environmental degradation appeared now to be irrefutable.

Still, there are important exacerbating features. By a tragic irony most of Africa's soil consists of infertile sand and laterite soils. It is rich in iron and aluminium compounds, and soon becomes rock hard on exposure to the sun and air. For this reason it is more than ever imperative with laterite soils to preserve vegetation. Unlike the rich humus soils of the temperate zones, laterite soils retain little water, and erode easily.

Hence over-grazing is suicidal for regions like Ethiopia and the Sudan. Soil erosion and desertification soon sets in, to cause uplands to lose, according to one UN estimate, one billion tonnes of topsoil each year. As a result the Sudanese president claimed that the Sahara has been encroaching upon his country at a rate of about three miles a year.

Soon the dreaded vicious circle sets in. Even the sparse rainfall is ultimately lost to the soil. Evidence has shown that deforestation in Latin America has seriously affected local and regional climatic patterns. Soil erosion frequently results when trees and vegetation are removed. In the long run desertification will increase at the expense of good soil that will sustain a fertile agricultural economy. Now some 20% of Latin America is affected by aridity of varying degrees of severity. Particularly badly hit are Chile, Mexico, Argentina and Peru.

But the vicious circle is sustained in other, more intricate ways. The Earth's albedo (or reflectivity) varies automatically with the seasons, as the ice sheets advance and retreat. But man is interfering with the albedo on an ever-greater scale with his glass and concrete cities and his highways. Deforestation in addition involves a change in albedo, because the canopy of trees, stretching for hundreds of miles in places like Amazonia, is generally of a darker hue than the surrounding terrain. Felling trees, then, means that more sunlight will be bounced back into space.

The effects, however, of such a highly marginal reduction in solar energy will hardly be noticed by the population. It would represent a fall of a mere 0.5% in terrestrial heat in the worst affected areas. Instead a more noticeable impact is made on rainfall levels because of changes in air circulation and weather patterns.

More significantly there is a loss of some transpiration (the exuding of water vapour), simply because there is less vegetation around. Things would get evened out over a deforested area of say a few thousand hectares or so, since the shifting winds of the global weather machine will still be able to transfer moist air from vegetated to desert areas. The layman may be surprised to learn, in fact, that the warm air over desert regions contains as much moisture as that of fertile northern Europe. The problem is that the sun over desert regions burns so fiercely it can actually vapourize any tiny clouds that do appear.

Even so, as the desert spreads the moisture content of the air drops marginally. And if this drier air is not dispersed by the prevailing winds it will ultimately make it virtually impossible to halt the worsening aridity of the area.

The drying out phenomenon quickly deteriorates. Vicious storms whip up the dust and hurl it across vast distances. This is not a new phenomenon. Darwin noticed it when he first arrived in the Cape Verde Islands off western Africa. 'The dust falls in such quantities as to dirty everything on board', he wrote in the *Voyage of the Beagle*. 'Vessels have even run on shore owing to the obscurity of the atmosphere'. Darwin was shocked to realize the global dimension of African dust storms. When a scientific colleague analyzed the dust, he discovered that much of it consisted of the pulverized remains of South American animals. Recent research at the University of Miami has confirmed how this African dust storm plume moves with the seasons between Africa and South America.

Americans are only too familiar with the rapacity of giant dust storms. The Dust Bowl of the 1930s taught the authorities a painful lesson in land conservation. The World Almanac and Book of Facts of 1935 described 1934's disastrous weather like this:

'Drought and dust storms in the mid-West are destroying winter wheat at the rate of one million bushels a day... Corn planting is held back... cattle are suffering. It is estimated that 300 million tons of topsoil were blown away. It darkened the air to the Atlantic seaboard'.

Indeed, Reid Bryson, a controversial and well known professor of Meteorology at the University of Wisconsin, argues that manmade dust explains most weather abnormalities. Human influences on the weather, believes Bryson, are now much more important than that other rogue phenomena, the outgassing effect of volcanoes.

Do Volcanoes Cool?

As surprising as it may seem, every living thing on Earth owes its very existence to the planet's volcanoes. Responsible for the formation of the primeval seas and oceans, they continue to exude water vapour in vast quantities. And yet volcanoes are universally associated not with water but with fearsome and violent heat, and with the destruction of life.

Nowadays they are known to possess other unpleasant characteristics. They eject prodigious quantities of dust and sulphur into the air up to 25 miles high, well beyond what is known as the troposphere. In such events the particles fail to get washed back to Earth by the rain, and form dust veils that can literally circle the globe for years.

Ever since Benjamin Franklin suggested that the cold European winter of 1783–4 was the consequence of massive volcanic eruptions in Iceland there has been a vigorous debate over the connection between volcanoes and the weather. And the second half of the 20th century, with highly disturbed weather patterns, has seen a lot of volcanic activity: Mount Spurr in Alaska in 1953, Mount Bezymiannyi in Kamchatka in 1956, Mount Agung in Bali in 1963 and Mount St Helens in Washington in 1980 (a much filmed and analyzed event).

The cooling theory holds that plumes of volcanic vitreol can block out the Sun's light and heat. This theory in turn hangs on the tenuous historical evidence of the Franklinesque kind that says that in olden days giant volcanoes blew their tops with such violence that the world's weather was drastically altered for months at a time.

A leading exponent of the volcanic cooling school is Hubert Lamb, Britain's foremost climatologist and formerly a professor at the prestigious climatology unit of the University of East Anglia. He estimates that after Krakatoa erupted in 1883 some 25% of the Sun's heat was absorbed by the millions of tons of ash that poured into the lower atmosphere. Krakatoa was the most lethal eruption in history, killing 36,000 people in coastal villages, and waking others in Australia 2,000 miles away.

Lamb also points to the April 1815 Indonesian volcano Tambora which blasted so many millions of tons of pulverized rock into the sky that the island of Madura, 500 miles away, was plunged into darkness for three days. Bizarre weather conditions were said to be experienced in the northern hemisphere, with snow in the following June.

But the volcanic cooling theory begs a lot of questions. It can explain historical climates much better than contemporary ones. We must remember that Lamb's reputation is derived from his scholarly and theoretical work rather than from his experimental atmospheric research. Constantly pouring over ancient weather archives and ships' logs, he seizes on any reference to unusual weather in order to compile glutinously detailed and complex volumes on the history of the world's climate. He is credited with devising a 'dust veil index', with which to compare the blackout properties of differing volcanic blasts.

The problem with this approach is that statistical relationships can easily be confused with causal ones. If tables of figures are stared at for long enough, a pattern seems to emerge. Lamb once believed he had detected a centennial cycle in the weather. The storms at sea and severe winters of the 80s of the last centuries (1680, 1780, 1880 and so on) implied that the present 80s might not escape the trend.

One difficulty is that the two most recent spectacular eruptions, Mount St. Helens and El Chichon, didn't confirm the cooling theory. Computer climate models of simulated Mt. St. Helens dust could only come up with a mere 0.1C cooling at the North Pole, a figure easily brought about by natural climatic factors. The Mexican El Chichon blast of March 1982 was eagerly predicted to cool the Northern Hemisphere for years to come. And yet after El Chichon Europe enjoyed two very warm summers in a row.

Furthermore, allusions to historic volcanic eruptions suffer from the fact that scientific techniques for ascertaining the actual nature of the outgassed materials were not as highly developed as they are today. Even now, as Lamb admitted in a recent scientific article, the varying amounts of water vapour, carbon dioxide and sulphur oxides differ 'from volcano to volcano, and from case to case'.

However, it is known that volcanoes usually emit 64% water vapour, 24% carbon dioxide and 10% sulphur, with a 1½% trace of nitrogen. *These are all Greenhouse gases*. Furthermore, the sulphur particles can combine with nitrous oxides already present in the atmosphere from fertilizer use. If so, volcanoes could contribute to the acid rain problem. In addition we need assurances that hot volcanic aerosols do not have the same radiation-absorbing characteristics of some lighter pollution particles.

But the most serious criticism of the 'volcanoes-as-climate-shapers' theory is that explosive eruptions always occur at random and often without warning. Their climatic impact is similarly random, often localized, and invariably short-lived.

The argument that volcanoes can *warm* the atmosphere was voiced in 1982 by F. Arnold and T. Buhrke of the Max Planck Institute of Nuclear Physics at Heidelberg. Taking advantage of the El Chichon dust cloud with data gathered from stratospheric balloons, Arnold and Buhrke calculated that the stratosphere warmed up slightly because the acid particles absorbed sunlight and *redistributed* it.

In any event, many scientists still testing the volcanic theories believe the cooling of the atmosphere they might cause must either be temporary or inconsequential.

Ronald Gilliland is an astronomer who has made a special study of the Sun's behaviour dating back to the 17th century. His findings reveal an intriguing solar pulsation that seems to take place every 76 years. Gilliland has worked with Stephen Schneider, one of America's top climatologists working at the US National Centre for Atmospheric Research, and whose name will crop up frequently in the rest of this book. He, too, has extensively examined the volcanic cooling theory. Together Gilliland and Schneider have taken the best solar and volcano data to work out the combined influence on world temperatures over the past 100 years. Then, in conjunction with Tom Wigley and Mick Kelly of East Anglia University they subtracted the volcanic and solar effects from the figures to reveal a distinct 1C rise in temperature — proving the Greenhouse Effect was real and irresistable.

In the meantime scientists were becoming worried not only about recent weather trends but about their own vacillation on the subject. Books and articles by climate experts argue with equal force for and against a warming. Some writers, like Lowell Ponte, maintain that the Earth is both warming and cooling at the same time.

While the scientists pondered, black storm clouds, as if an omen, suddenly precipitated their load in September 1984 as physicists attended Britain's annual British Association meeting. Professor Lamb reflected the prevailing sense of bewilderment in his title address: 'The Future of the Earth — Greenhouse Earth or Refrigeration?' Plaintively, if unhelpfully, he told the Association that 'changes that menace the entire world economy may be before us'.

Much of this uncertainty stems from the enormous complexity of the web of climatic cause and effect. Climatology is discussed within the framework of certain dogmas and laws that succeed in making even short-range predictions a tricky business. Most of them are engineering analogies. One is the flywheel analogy which decrees that the oceans act as giant heat reservoirs. They take a long time to warm up, but at the same time take a long time to cool down. They could still bathe the world in warm, moist air even at the predicted beginning of a new Ice Age.

Another engineering analogy is the feedback mechanism, Nature's own answer to the in-built governor. Like an illness that sets a person's antibodies working overtime, things start to go into reverse. The example often quoted is the halting of the glaciers in their tracks and the consequent shrinking of the area of reflective whiteness, which in turn enables the land below to absorb more heat, and so on.

In the meantime the role of the oceans is still capable of baffling the most knowledgeable climatologist, and the most sophisticated computer. For example, more solar radiation can cause more evaporation from the seas and hence create more clouds. But this could either block out the Sun's rays, or turn the world into a Turkish bath.

A recent criticism of the Greenhouse Effect theories focused largely on this controversial aspect of the issue. Richard Somerville, a meteorologist with the Scripps Institute of Oceanography, argued in April 1985 that denser clouds, as a result of the warming, will reflect more sunlight. Hence the warming will only be half as great as the best climate models now predict. He even repudiated the theme of this book by suggesting that ocean levels would fall instead of rising. But it is not just solar penetration we are worried about. The anthropogenic heat-up is occurring *below* the clouds, which the clouds themselves will trap. Molecules of water, after all, are *even better* absorbers of infrared heat than carbon dioxide.

There is still endless speculation about the albedo factor. Is Man making the Earth turn a lighter shade with his deforestation and his roads and cities? Or is the Greenhouse Effect more than making up for any cooling that this might bring about? Professor Richard West, at a meeting of geologists gathered for the British Association annual jamboree, asked whether environmental pollution would alter the course of climatic trends. He raised ominous questions about CO^2, sulphur dioxide and nitrogen oxide. There are no easy answers.

The Evidence of the Warming

However, the pro-Greenhouse school, on balance, held sway. Whatever might be the natural, cyclical, trend for the world's climate, it was becoming clear that the warming must be having some sort of impact, even if only in the short term. The trouble is that the short term may turn out to be rather more prolonged than many anticipate. A recent Pelican book by John Gribbin, a distinguished and prolific interpreter of climatic issues for the layman, argued for a warming first, to be followed by a return to Ice Age conditions a little later, say in a few hundred years time.

The issue of Man versus Nature raises its head again. The anthropogenic warming may be so unnatural and so unremitting that even a full-blown Ice Age may have a hard time trying to reverse events. Stephen Schneider calls the effect a 'creeping, inexorable problem'.

We have seen in the last chapter how carbon dioxide levels have been rising ever since the smokestack industries began operating at full blast in the mid 19th century. It is not without significance, then, that the East Anglian Climatology Unit, after extending the climate record back to 1851, can show that the world has been warming up ever since, with only minor setbacks. One major reversal seems to have been the immediate post-war period; roughly the twenty-year stretch from about 1950 to 1970. Then there seemed to be a distinct cooling, with British holidaymakers nostalgically yearning for the balmy summers of yesteryear.

John Gribbin, however, is at pains to put the pre-war warming into historical perspective, and he attributes it largely to unusual solar factors. The present erratic weather may again be due to changes in solar radiation. In any event we should not, he says, take the first half of the 20th century — exemplified not so much by global warmth as temperate equability — as the norm: 'Our standards of normality are based on the most abnormal conditions of the past 1,000 years'.

Certainly Gribbin is a reluctant supporter of the Greenhouse theories, and tends to place more emphasis on astrophysical explanations for climatic change (see Chapter 7). Gribbin, however, is in an invidious position. He is one of that new breed of scientists in Britain and America who have struggled to get the meteorological establishment to take the study of climatic change seriously. Hitherto government-trained weathermen have refused to read anything significant into a run of exceptionally cold winters or warm summers. 'The weather is like that', they say. 'It all averages out in the long run'. Gribbin would be the first to point out, though, that the average itself is on the rising gradient.

Chapter Six has shown the successes the climate modelers have already had in measuring the gases in the atmosphere, and with comparing present temperature scales with historical ones. The roll-call of scientific institutions producing concrete evidence of rising temperatures in the sea and in the air is growing annually. For example, Graham Farmer and his colleagues at the Norwich-based Climatology Unit believe that the chilly 1960s and 1970s are now part of history. The year 1984, they claim, was the 5th warm year in a row, with 1981 being the hottest ever recorded in the northern hemisphere.

The Unit, funded by the US Department of Energy to look into the CO_2 effect, has produced figures showing that since 1970 not only has a cooling phase since the War been reversed, but, as we have seen, there is now greater variability in the world's weather.

The EPA, in their 1983 report, *Projecting Future Sea Level Rise*, say that if one goes back two million years, the climate has *never* been more than 2 to 3C warmer than it is today. In the last 100,000 years it has been at most 1C warmer, and in the last 1,000 'at most 0.5C warmer'. Earth has warmed 0.4C in the last century alone, and another 1C can be added to the temperatures of this century. Indeed, a sudden acceleration of the heat-up took place in the seventies — just as European and American winters started to misbehave and the Saharan drought intensified.

What is more revealing is that the oceans, even with all their in-built inertia, are warming perceptibly. The North Sea is up a phenomenal 4F since the war. It is changing the types of marine vegetation. The further north you go the greater is the rate of change. In some parts of the Arctic the temperature has risen by as much as 20F since the beginning of the century. Many Arctic ports are ice free several months of the year longer than they used to be. Furthermore, on a recent cruise in the North Pacific, oceanographers discovered that plankton are blooming as a result of the CO^2 build-up to such an extent that two to three times more organic matter is being produced by photosynthesis.

Deforestation goes hand in hand with the warming, and not just because of the reasons spelt out in Chapter 4. A warmer climate can kill off the shallow roots of trees. Indeed a Soviet magazine article recently attributed the rise in carbon dioxide and tree felling to the thawing out of the permafrost in the Zeya region, just north of the Chinese border.

Arthur Lochenbruch and Vaughn Marshall, from the US Geological Survey, in their study of the Arctic permafrost region in 1986, also noted a conspicuous disturbance of Earth's temperatures. Measuring temperatures to depths of 600 metres over 100,000 square kilometres they discovered that the permafrost surface has become warmer by two to three degrees centigrade in just 100 years. Particularly revealing is what is happening to inhabited areas in Siberia and Alaska. Buildings erected on what for generations have been solid foundations are beginning to topple over. At one costly US government building the ice on one side of a rock-hard surface had all evaporated. Discounting speculation that the warming trend ought to be highly beneficial to Russian agriculture, Soviet scientists are instead worried that lasting harm could be done in Siberia as areas of permafrost become marsh, especially in summer. There are fears for road and rail links, and construction works, which could all be disrupted. As it is, special heating pipes are placed well above ground to prevent any warming of the permafrost and the subsequent sinking of buildings.

New insect pests attracted by the warmer climate now pose a serious threat to the great deciduous forests of Quebec. Experts at the forest pathology laboratory at Canada's Laval University have concluded that at least one half of all the country's birch trees have been destroyed over the past 40 years by rising temperatures.

In addition Canadian Great Lakes are higher than normal, and snow for the first time is melting on mountaintops. The dwindling band of Eskimoes in Canada are being left destitute as the caribou continue to move to the cooler north. Many newer types of trees — such as willow, spruce and maple — are now growing in northern areas such as Sweden, Finland and Alaska. Wheat is being grown in Canada at a more northerly latitude than before.

Americans are now experiencing far-reaching climatic changes. No longer do prewar temperature gradients prevail. Temperatures are rising everywhere. Cleveland, Ohio, now has an average winter temperature more appropriate to prewar Cincinnati. Cincinnati in turn can now boast winter thermometer readings of 35.5F which are more typical of Washington, while the Washington average for January (1985 was a rare exception) has risen to 39F.

Future Weather

In the meantime the East Anglian Climatology Unit envisages wetter and warmer autumns. Winters will also be colder and drier, and we could have warmer and drier summers. This theory puts paid to speculation about a new Ice Age. More warm summers — regardless of how cold the winters get — means that the winter ice and snow will be melted before it has a chance to linger on the ground and creep further southward (in any event it is not exceptionally cold winters that make the ice sheets grow, but somewhat milder, and more snowy, winters).

On the other hand permanently warm summers could cause major disruption to the production of British root and grain crops. Later, as the CO_2 effect became more pronounced, the summers in the north would become hotter. The winters, although cold, would be shorter as the warm springs began to take hold earlier. Parts of the fertile prairielands in America would become as arid as they were in the 1930s. The world's arable wheatlands would be pushed northwards to the benefit of Russia and Canada.

The Unit's conclusions were supported by forecasts made by the National Research Council, in its own report on the Greenhouse Effect. They bleakly prophesied widespread weather disruption in the future, with rainfall and temperature patterns being utterly altered. John Hoffman, of the EPA, also believes that statistically significant changes 'are likely to be here by the years 1990 to 2000'. Rainfall patterns would be changed, he agreed, irrigating some existing deserts and bringing drought to other fertile regions. 'The projected warming for the next century', concludes Hoffman, 'would be 10 times as rapid as the historical trend'.

H.W. Bernard, summing up in his popular American version of *The Greenhouse Effect*, says that within the next decade, by the year 2040, 'our climatic temperature may soar to beyond anything known within the past 125,000 years. The average hemispheric could be as much as 5F what it is now'. These temperatures, says Bernard, would match those prevailing between 4,000 and 8,000 years ago, when the ice sheets of the world began to melt catastrophically.

This last statement should make us think. The warming of those ancient times was due to changes in Earth's orbit around the Sun. Possibly it had something to do with changes within the Sun itself.

Could the Greenhouse theories, after all, be hopelessly wrong? We are, in the view of one late eminent scientist, heading right for the middle of what he called 'the Great Summer' for quite natural astronomical reasons. If so, there is nothing Mankind can do about it — except possibly make it worse for ourselves.

Chapter Seven:

THE COSMIC CONNECTION

Any writer arguing for a global heat death will have irrefutable scientific authority to back him up. For there can be no doubt at all that the Sun will one day scorch the Earth to a cinder. As the Sun's hydrogen fuel becomes depleted, and its core shrinks, its temperature will peak. Soon after it will become a Red Giant, and suddenly become 1,500 times as bright as it is now. The melting of the polar ice caps will become a trivial matter. The entire Earth will be vapourized.

Fortunately, however, such an event is scheduled for 4,500 million years hence, and there will be plenty of warning of the impending catastrophe. But the foregoing scenario proves how vitally important is the behaviour of the Sun to human wellbeing, indeed to the wellbeing of the entire biosphere.

We must bear in mind that throughout Earth's history many remarkable things have happened to it on its relentless journey around the Sun. There have been periods of rotational acceleration arising from axis wobble or tilt. The Earth, too, has frequently adjusted the distribution of its mass, and it may have passed through clouds of interstellar dust. Many scientists believe that these phenomena have had drastic and even sudden repercussions on our climate. Stephen Schneider, for example, believed that about 90,000 years ago a shift to near glacial conditions as a result of orbital events occurred in less than a century.

Curiously, the myths and legends of ancient civilizations acknowledge the vulnerability of the Earth to cosmic and solar events. People have, for millenia, worshipped the Sun as a god, realising how dependent they were on its existence for light and warmth, and for the fertility of the Earth. Sun worship is at the root of most religious festivals. But, like other gods, the Sun could be

Janus-faced, and easily stirred to anger unless appeased.

In the formation of early catastrophe theories a wildly careering Sun was blamed for causing countless Earthly disasters. Plato's writings were full of warnings of periodic annihilations of the Earth by fire and flood. In one Plato story Phaeton harnessed his father's (the Sun's) chariot, but lost control of it, to bring about Earthly doom. 'Truth', wrote Plato, 'lies in the fact that the heavenly bodies moving around the world deviate from their courses, and after a long time the Earth is scorched and consumed in huge fires'.

Egyptian priests told Herodotus that 11,000 years before the axis of the Earth became displaced, 'the Sun had removed from his proper course four times and had risen where he now setteth and set where he now riseth'. Old Chinese records speak of a time when the sky suddenly began to fall northward. The solar system was left in chaos as the planets flew apart.

The seeming catastrophic power of comets, probably because they were clearly visible to the naked eye as missiles from outer space, have fascinated learned men through the ages. In 1696 one William Whiston published a doomsday book arguing that in Biblical times a huge comet flew past the Earth to create massive tidal waves and landslips, to shatter mountains and flood the Earth up to six miles deep. Edmund Halley, who gave his name to a famous comet in the 17th century, implied that Earth must have been constantly bombarded by comets.

Many scientists of that era, in the wake of Halley's Comet, had their God-inspired catastrophism reinforced. In the world of the Marquis de Laplace (1749–1827), a genius in mathematical astronomy, the Earth was subject to unpredictable cosmic forces, including heavenly collisions. In his *Exposition du Systeme du Monde* he used two pages to argue that mankind should learn to accept potential cosmic dangers: '... entire species would be annihilated; all monuments of human industry overthrown; a great portion of the human race and animals would be drowned in the universal deluge'.

Eventually, as the science of astronomy made progress, celestial collision theories became the domain of the eccentric iconoclast. It was left to neo-scientists and occultists of the Atlantean variety to keep the catastrophic heritage alive with intriguing speculations. During this century the works of Emmanual Velikovsky, offering an astrophysical interpretation of the history of the solar system now known to be hopelessly wrong, nevertheless received worldwide publicity.

He came to the conclusion that a massive comet must have appeared in the skies in the third millenium BC. This, he surmised, was the time of the Biblical Exodus. After roaring past the Earth, causing the planet to tilt and bringing about weird atmospheric disturbances, the north polar ice fields were shifted further north. Enormous tidal forces hurled entire oceans over land masses, and caused massive earthquakes and volcanic eruptions. The sea evaporated, mountain ranges collapsed, and even the solar planets were bounced from one orbit to another. In his book *Earth in Upheaval* Velikovsky writes: 'Tidal waves traversed continents, moving by inertia when the daily rotation of the Earth was disturbed... These tidal waves, augmented by others produced by the extraneous fields of force generated by boulders, distributed marine sediment over the land... and floods (were) caused by the melting ice cover...'

Now, however, catastrophism — based more soundly on astrophysics — may be making a comeback following the publication on both sides of the Atlantic of new 'impact extinction' theories. For example, scientists James Lovelock and Michael Allaby argue, in their book *The Great Extinction* (1983), that the dinosaurs might have been killed off by a kind of cosmic winter. A comet or planetismal entering the atmosphere at a particularly shallow angle, could have ricochetted off to disperse thousands of tons of material. This would then have been converted into aerosol particles by the heat of entry. Volcanoes could have erupted with tidal waves sweeping the coastal areas. But it was probably the dust in the upper atmosphere that caused the most harm to Earth by distorting the climate. Louis and Walter Alvarez, of the University of Berkeley, California, developed a similar theory in 1980.

In their acclaimed 1982 book, *The Cosmic Serpent*, two distinguished Edinburgh astronomers, Victor Clube and Bill Napier, maintain that theories about Earth being struck by celestial bodies 'have not been taken seriously... one can open almost any textbook on palaeontology or geology to find the evolution of Earth discussed as if the planet existed in isolation'.

They point to the many thousands of meteors that have been

spotted careening through the skies in recent years. They suggest that a giant 'comet or asteroid' struck the Earth at some distant time in the past, to generate enormous kinetic energy. This event brought about violent episodes of mountain building, volcanic eruptions, sea-level changes, 'climatic excursions' and magnetic field reversals. The pulverising effects of such a catastrophic impact would lift so much dust into the air that it would blot out the Sun's heat 'several times over'. Meteorites, or larger bodies, by this reckoning, 'might have caused' the Biblical deluge.

It is now more acceptable to invoke planetismals and meteoritic missiles as an explanation for Earth's 'mysteries'. The credibility of such theories largely rests on a spectacular and much researched event that occurred in Tunguska, Siberia, in 1908. In spite of the more esoteric speculations about anti-matter, scientists are virtually unanimous in their opinion that what devastated thousands of square miles of Siberian forests in the summer of that year was a careening solid object from space.

There was an explosion high in the atmosphere, and a shower of debris rained down from the skies. The pressure wave was said to have knocked down a man 35 miles away. The searing heat of the meteorite (or comet) suddenly raised the air temperature by some hundreds of degrees. The air's nitrogen and oxygen molecules became dissociated, some of which combined to form oxides of nitrogen. And as the shock waves of the impact spread outwards, vast quantities of nitric oxide were generated. According to Richard Turco, writing in the science magazine *Icarus*, this gas would have disturbed the chemical balance of the upper atmosphere. The ozone in the atmosphere would soon have been depleted by up to 45%, to leave the planet highly vulnerable to harmfully increased doses of solar radiation.

The veteran astronomer, Sir Fred Hoyle, once voiced a similar Doomsday scenario. Instead of the Earth cooling under a dust blanket, water vapour high in the upper atmosphere would freeze to form a vast belt of tiny ice crystals — 'diamond dust'. This glittering veil would then lock an Ice Age into place for thousands of years by reflecting sunlight back into space. The Earth is left void and frozen until the ice crystals finally disintegrate.

The Unsteady Sun
Scientists agree, then, that Earth is still at the mercy of events in

space. It is not without significance that many climatologists are astrophysicists, or work closely with them. Many of them focus not on the danger of global floods or Ice Ages arising from comets churning up the oceans, but from flickering disturbances occurring on the surface of the Sun.

But if the Sun flickers, does this mean that it is not 'constant' as scientific dogmas decree that it is? Some astrophysicists, like Paul Davies of Newcastle University, argue that to all intents and purposes it is, but at the same time it is in a state of 'restless torment'. It was students of the Sun who first discovered that solar radiation fluctuates. It throbs, pulsates, shoots off jets of hydrogen gas, and, as if catching a chill, develops sunspots. It is the seemingly tiny variations in the Sun's output, of which the man in the street would be totally unaware, that can have widespread repercussions on the terrestrial climate, and even on the height of sea levels.

The Sun, like the Earth, has a built-in heat stabilizer. Like all stars, the Sun generates heat and light in its core by a nuclear reaction in which atoms of the lightest element, hydrogen, combine to form atoms of the next lightest, helium. Though the core of the Sun is fantastically dense and hot, the sunlight that we see emanates from the surrounding atomic structure, which is released and reabsorbed millions of times. The shining surface of the Sun, the photosphere, has a temperature of about 6,000C — roughly the same as that reached in arc welding. Yet it takes so long for the light to escape from the Sun, that the sunlight we now experience was generated in 8,000 BC.

Now, if the nuclear furnace at the core of the Sun suddenly produces more wave radiation its outer reaches would expand and radiate the additional heat into space. But the turbulence of the Sun manifests itself in giant flares and sunspots. A flare is a burst of energy, consisting mainly of protons, that sweeps across the solar system. The aurora borealis at the North Pole is a manifestation of these flares, as the protons interact with Earth's magnetic field.

The magnetic field, however, may momentarily fail, to allow the lethal protons to enter the atmosphere. Then dangerous levels of nitric oxide would be created, to ultimately deplete the ozone layer and thus cause temperatures, through complicated chemical and atmospheric processes, at the surface to rise prodigiously.

Fortunately this magnetic flip does not seem to occur very often; the last occurred about 700,000 years ago. It is all the more alarming, then, to be told by scientists that it may be about to flip direction quite soon. Subir Bannerjee, a geologist at the University of Minnesota, says that the magnetic field has halved its strength over the past 4,000 years in a prelude to a new and imminent reversal. As he reckons this would take a couple of thousand years to finally accomplish, we could expect some extreme and bizarre weather in the meantime.

There are other types of solar irregularity. The hydrogen of the Sun, in the process of nuclear conversion into helium, leaves behind trace elements of the heavier atoms such as carbon, magnesium and iron. These elements will ultimately make up the rocky and barren Black-Dwarf that the Sun will become in billions of years time when all its nuclear energy is expended. In the meantime — every time these trace elements are ejected from the fore — massive solar convulsions at the perimeter of the Sun and its atmosphere occur. And, according to Ernst Opik of the Armagh Observatory, it is these throbbing blasts of solar energy that cause Ice Ages.

Then there are the 'cosmic dust' theories of climatic change. W.H. McCrea, of Sussex University, suggested that the solar system is itself in orbit round the Milky Way galaxy. The journey takes 250 million years to complete. Earth, in the process, frequently passes through lanes of cosmic dust and debris in the star-filled spiral arms of the galaxy. The dust gets sandwiched between the Sun and Earth, reducing the sunlight reaching us for periods lasting years or more. The solar system emerged from a dusty strip of the spiral arm 10,000 years ago, at a time when the last Ice Age was ending.

Others interpret the dust lane theory differently. The dust falls into the Sun, making it blaze more fiercely, and this brings on an Ice Age in accordance with the 'warm Sun/cool Earth' principle. Or perhaps convection in the Sun is inhibited by the dust lanes, only to flare up more powerfully later as the cloud moves on, rather like someone removing a lid from a pressure cooker.

Why extra solar heat should cause Ice Ages is one of those strange paradoxes peculiar to climatology. The conventional argument dictates that heat causes evaporation from the oceans, which can turn to snow in higher latitudes, which in turn can make the ice-sheets grow. The 'hot Sun/cool Earth' theory is now, in any event, losing favour with scientists. Indeed, the more logical theory is now preferred: a hotter than usual Sun will sear the Earth, any additional atmospheric moisture created in this way merely adding to the Greenhouse Effect.

The Sunspot Cycle

But it is the curious dark blotches on the Sun, known as sunspots, that stimulate the most scientific interest. They have been suspected for a long time to have some connection with the Sun's variable output. Some are massive — up to 10,000 miles in diameter — and can last up to a month. They seem to bubble up from a mighty expansion of gases emanating from the core of the Sun. And it is this expansion — using the same principle as that of the domestic refrigerator, that causes the temperature to drop.

One theory suggests the Sunspots are triggered into life by the influence of other planets in the Solar System. For example, when Earth and Venus are aligned on the same side of the Sun the number of sunspots are said to differ from their normal pattern. Recently this belief has become known as the 'Jupiter Effect'. This theory dictates that certain planetary alignments exert a gravitational pull on the Sun, which can trigger sunspot activity, which can in turn trigger geological disturbances on Earth, including Earthquakes. The original authors of *The Jupiter Effect* (1974), John Gribbin and Stephen Plagemann, have recently placed less emphasis on the geophysical repercussions, but they maintain that the Effect can still discernably alter terrestrial weather patterns.

The most striking feature of the sunspots is that they have magnetic fields some 1,000 times stronger than both the field of the Earth and the rest of the Sun. Herein lies one possible clue to climatic change on Earth, since sunspot magnetism probably interacts with Earth's magnetic field to bring about magneto-meteorological effects.

The validity of the sunspot theory rests on both historical and geophysical records. The most often cited illustration is the strangely cool period lasting from 1645 to 1715 AD, known as the Little Ice Age. During this time when temperatures were abnormally low, no sunspots at all were recorded. From that time onwards sunspot appearances and terrestrial temperatures have

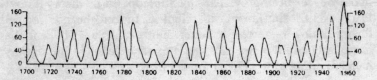

Over the centuries the changing level of sunspot activity bears a close relationship to changing levels of solar activity. A cooler Sun seems to correspond with a minimal number of sunspots.

Source: John Gribbin, Future Weather, Pelican, 1983

been linked, most particularly in recent times when the Earth's giant polar ice caps have been examined. Edward Zeller of the Geophysics Dept of Kansas University, for example, has traced billions of years of compacted snowfalls containing nitrogen compounds. It could only have been solar or cosmic events that could have dissociated nitrogen particles from the atmosphere, so nitrogen-dating techniques can now match up ice core readings with historical weather recordings.

However, scientists are still not entirely certain that sunspots warm the Earth. Periods of inclement weather have also coincided with sunspots and solar flares. One complexity arises from the magnetic effects of the sunspots, so the knock-on effect on Earth depends largely on the current state of geomagnetic play — whether Earth's field is increasing or decreasing, and in what way, and in what hemispheres. Unfortunately the field tends to vary from one part of the world to the other, and from one epoch to another. Sunspots do seem to coincide with periods of global warming wherever Earth's field has weakened, or if not particularly pronounced, or has drifted somehow out of alignment. And this seems to be the usual state of affairs.

Much of the confusion on the subject perhaps arises from the Sun's own orbital spin. This would affect solar variability as much, if not more so, than sunspot activity, and indeed would largely override any effects arising from the latter. For example there is some evidence that the Sun's spin at its equator sped up some five percent during the Little Ice Age, all the while carrying sunspots with it.

Sunspots, after all, are just one type of fluctuation that the Sun seems to permanently undergo. Indeed the Sun often shows signs of chronic instability, so much so that scientists now have to be

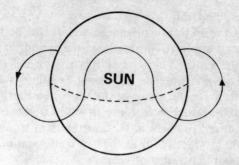

How the Sun's magnetic field wavers across its rotational axis.

Source: John Gribbin, The Climatic Threat, Fontana, 1978

careful about writing about a 'wobbling' or 'fading' Sun for fear of alarming the public, as one of Britain's daily tabloids did recently.

The kind of regular variations in radiation it undergoes are similar to those stars known as cepheids, which fluctuate periodically in a way that depends on their mass and luminosity. As the Sun is plasma, all of its energy received at the Earth's surface is electromagnetic, as we have seen. So one form of power can decrease while other forms can increase. The Sun, of course, emits direct heat and light. But it also radiates X-Rays, ultraviolet light and radio waves. It gives off energy in the form of particles of matter ejected from its outer atmosphere (or corona), known as the 'solar wind'. The two types of solar energy are best thought of as 'wave' and 'particle'. Hence if this latter type of radiation is suddenly boosted by the internal mechanics of the Sun — a solar flare erupts, for example — this doesn't mean that the Sun suddenly gets hotter. Yet such an event would still have repercussions on Earth's 'weather machine'.

Some scientists believe the Sun fluctuates, like a thermostat, in response to conditions in the surrounding galaxy. Ronald Gilliland, of NCAR in Boulder, Colorado, says the Sun appears to 'breathe' in and out every 76 years, as confirmed by historical records of the Sun's diameter. What is interesting is that Gilliland says the Sun was at its biggest in 1911, and at its smallest in 1949. It should, he said, be at its biggest again in 1987. Then all kinds of energy, including infrared heat, will be at its maximum, thus adding to the Greenhouse Effect.

Earth's Changing Orbit

At first sight, the orbital theories seem to offer the best solution to the waxing and waning of the Ice Ages, the periodic warming and cooling of the Earth. They are simply the logical extension of what the layman understands about weather and seasonal variations: they are the result of the tilt in the Earth's axis as it circles round the Sun. Hence, if the tilt or the orbit becomes rather more pronounced than usual then extremes of temperature will arise, and will last for as long as the new configuration lasts — for perhaps tens of thousands of years.

It was in 1941 that the late Milutin Milankovitch, a painstaking Serbian mathematician, published his figures showing that the Earth's orbit occasionally became greatly stretched during the Earth's Ice Ages. Indeed Milankovitch, reflecting the seasonal analogy, actually used the words 'Great Winter' to describe the full Ice Age, and a 'Great Summer' to represent the interglacial.

The theory emphasizes the elliptical nature of Earth's orbit. This means that the Earth is not exactly the familiar 93 million miles away, but varies between 91.3m and 94.4m miles twice a year. When the Sun is closest to Earth (known as the perihelion) the northern hemisphere is deep in winter, and this explains why northern winters are milder than they would be if the orbit was precisely circular.

The axis tilt of the Earth also is not constant, as the direction of the tilt performs a full circle in space, once every 25,000 years or so. In addition the Earth travels round the Sun, slowly shifting its perihelion point, and gradually nodding from being more upright to less upright, over a cycle of about 41,000 years. It is this tilt that is the key, Milankovitch believed, to the Ice Ages. The more upright the tilt the more Earth will receive more evenly distributed solar radiation, meaning cooler summers and milder winters in the north and south — ideal for building up the ice caps. If one adds to this Milankovitch's idea of an increasingly eccentric orbit these 'Great Seasons' take 100,000 years to do a full cycle.

Hence, if Milankovitch's theories are correct, we are still in the 'Great Summer' of maximum temperatures. We will shortly, within a thousand years or so, be heading for the 'Great Autumn' which has led climatologists to speculate that the next Ice Age 'is just around the corner'.

Geological investigations seem to confirm the Milankovitch

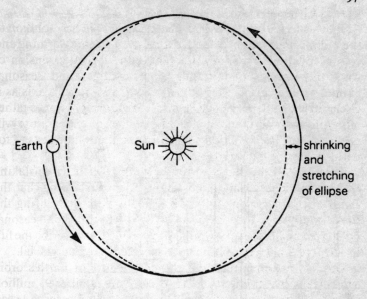

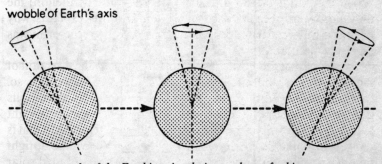

Change in angle of the Earth's axis relative to plane of orbit

models. The most famous of these was undertaken in 1976 by Jim Hays, John Imbrie and Nicholas Shackleton, all of Columbia University. They examined the nature of fossil shells dragged from ocean beds, and the amount of oxygen-18 molecules in them. They then wrote up a definitive account of the historic time span of advancing and retreating glaciers, and their bearing on the rise and fall of sea levels. Their findings seemed to confirm the three major cycles of approximately 24,000 years, 42,000 and 100,000 years.

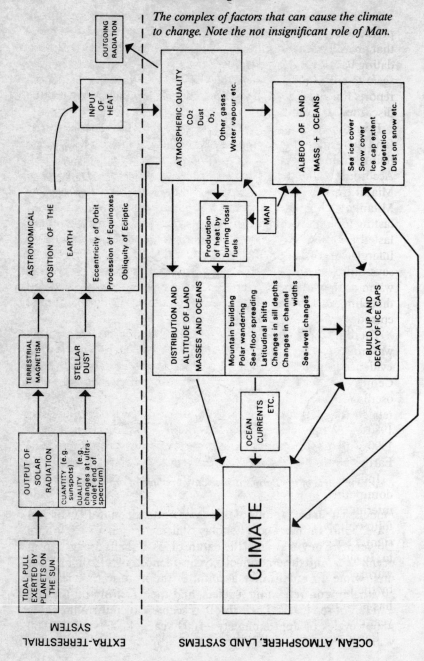

The complex of factors that can cause the climate to change. Note the not insignificant role of Man.

Furthermore an international research project into climate prediction, also headed by Jim Hays, has analysed seabed cores that provide valuable clues to past temperature changes on Earth, dating back over the last 450,000 years. These changes do, in fact, seem to correspond to variations in Earth's orbit. 'The evidence', reports Jim Hays, 'is so strong that other explanations must now be discarded or modified'.

Even those scientists who disagree with the Milankovitch cycles (and there are not many) still work on the assumption that we are living through a period of maximum terrestrial temperatures. According to the distinguished climatologist Cesare Emiliani, of Miami University, the Milankovitch time scales are too long. The Milankovitch model decrees long periods of warmth and cold lasting for equal periods, whereas Emiliani says that the Ice Ages lasted for about 200,000 to 300,000 years, interspersed with short interglacials of about 12,000 years. Over the last 700,000 years, he says, only something like 5% of the time have temperatures been as warm as they are now. Many scientists now agree with Emiliani that climates as warm as the present are 'short, wholly exceptional episodes'.

There are other theories to support the idea of a present, natural, warming. A Danish team headed by Wilf Dansgaard found evidence from ice drilled from the Greenland ice sheets at Camp Century of much smaller warming and cooling rhythms. These oscillations occur at 80 and 180 year intervals, and may, too, be related to solar variations. Furthermore we may — in the late 1980s — be right at the peak of one of these warming trends. After 1990, with the Greenhouse Effect overlaying this natural trend, the Earth's climate could really start to soar.

Indeed, the man-made warming could ultimately win out, dominating the natural cycles. The 80 year warm and cold interludes may instead turn out to be 40 years of rapid warming followed by a similar period of moderate warming, with hardly significant cooling of the type the Earth has been historically used to. In just 15 years from now the world could be heading for a warm climatic state that it has not experienced for 1,000 years. And by the middle of the next century it may become warmer than it has been for 8,000 years. And the century after that a 70 million-year climatic optimum may be reached.

Earth may never experience another Ice Age again.

Chapter Eight:

THE DISINTEGRATION OF ANTARCTICA

To most people Antarctica is — literally and figuratively — the Last Place on Earth. It is as remote and mysterious as the moon. It is sometimes so interminably cold down there in winter it can freeze the surrounding Southern Ocean solid for several hundred miles around. When this happens this ice-caked Continent becomes even more massive.

It is larger than Europe, and if it were centred on North Dakota it would extend from the Atlantic to the Pacific, and from Mexico to Northern Canada. It covers a total of 5½ million square miles at the bottom of the Earth, some 3% of the Earth's surface, more than two miles deep in places. This mammoth glacier, the world's highest continent, represents 85% of the 8 million or so cubic miles of the world's total snow and ice reserves.

Yet it is scenically beautiful. The striking mountain chains of the peninsular are reckoned to be a continuation of the South American Andes to which they are linked by a massive submarine arc reaching out to enclose the Scotia Sea and the Falkland Islands. Some of these ice-clad mountains form an awe-inspiring plateau rising to over 13,000 feet above sea level.

Few, however, have ever visited Antarctica. It is often literally unapproachable. The continent is surrounded by a formidable barrier of treacherous floating pack ice, navigable to shipping in a few select places only during the short summer season. Northwards of the pack ice lies Drake Passage — one of the world's stormiest stretches of ocean. Here the notorious trade winds such as the 'roaring forties' are a menace to shipping.

The major centres of population in the northern hemisphere are no less than 8,000 sea miles distant from this inhospitable world. The nearest mainland is the tip of South America, which is still 600 miles distant. The Antarctic Ocean is dotted with a few bleak but observationally strategic islands in a waste of ice and frozen water. Indeed it would appear that the peoples of the Earth are more affected by what happens in the Arctic, rather than at the other end of the world. With the exception of India and China the most populous regions are situated in the northern hemisphere — some 90%. Most of the world's sea and air routes also lie north of the equator, even crossing the Arctic itself. The vicious northerly winds that so often give Europe its 'cold snaps' also emanate from the Arctic.

The polar regions derive their name from Greek cosmography. The word *arktos* was given by the Greeks to the constellation which rotated about the north polar regions. Soon it came to mean the sea and land surrounding the North Pole. The southern pole, being opposite the Arctic, was called Antarctica.

The notion of a spherical Earth, first propounded by Pythagorus in the 6th century BC, became conventional wisdom. As soon as the human mind got used to the idea of the Earth as globe-shaped the existence of a southern continent seemed logical. Indeed, in the interests of symmetry it seemed necessary. Later, as the idea jelled, Claudius Ptolemy, in the second century AD, conceived of an immense southern land mass which he called 'Terra Incognita' (later to be known as Terra Australis). This was some sort of land bridge reckoned to be sited between Africa and the Malay peninsular.

Ptolemy's ideas were, at the time, sacrilegious, and languished for a long while, until the late 15th century stirrings of the Renaissance. Soon a great era of geographical exploration got under way, to be equalled only by discoveries made in the mid 20th century. We now know that, according to Continental Drift theory, Antarctica was Gondwanaland, and rifted apart from Australia to move slowly into position over the South Pole. Then the ice gradually built up to its present coverage.

The Pole Shift Theories
This new knowledge of a mobile and fluctuating ice continent seems to have fired the imagination of a generation of Doomsday writers. The South Pole was viewed as a giant gargoyle at one end

of a smooth sphere that was already tilted away from the upright. With a half-digested knowledge of Milankovitch cycles, earth 'wobbles' and Continental Drift theory, it was obvious to many neo-scientists that Antarctica could not forever build up mass without sooner or later flipping the Earth over.

And yet the end-result of such a 'polar flip' would not result in the destruction of the Earth. Instead it would result in a global Floodwave of unparalleled — if not unimaginable — proportions. The Biblical flood and the death of the dinosaurs are cited as real and catastrophic examples of such a pole shift. According to the late Hugh Auchincloss Brown, an amateur scientist who spent a lifetime formulating polar flip theories, air and water would continue to move through inertia, thus causing the seas to rush over the continents. John White, in his book *Pole Shift*, describes how an observer standing on Earth would see a giant tidal wave some 13 miles high rampaging towards him at hundreds of miles an hour!

Surprising support for the pole shift theories has been given by the British physicist Peter Warlow in his 1982 book *The Reversing Earth*. Like earlier writers, Warlow supports his theories with the narratives from ancient documents. If, he surmises, the Earth tilted so as to push China further southwards, there would be a corresponding movement of the star pattern in the opposite direction. Warlow refers to a Chinese legend which did in fact mention both phenomena.

The same tilt apparently brought about the opposite effect on the other side of the globe, pushing North America further towards the Arctic. This explains, wrote Warlow, why the Delaware Indians had a Noah-like flood legend set in freezing cold conditions. 'The Hopi Indian legend relates that the world turned to solid ice after the sea had sloshed over the land as a result of the world spinning around crazily and rolling over', he wrote.

Geophysicists, as one might expect, view the polar flip theories with extreme scepticism. The energy required to change the Earth's axis of rotation would be enormous. Shortly before Warlow's book appeared, an American physicist published some calculations deducing that the torque required to invert the Earth is 200 times greater than envisaged by Warlow. This would, it seems, be too large for any cosmic body to bring about.

The major problem with this type of catastrophe theory is one of energy. The Earth is like a giant gyroscope with terrific and irresistible momentum. This is not to say that the pole shift theories have no scientific credibility. It depends, largely, on whether they are placed within the catastrophic or uniformitarian frames of reference.

The difference between the two depends on one's understanding of geologic time. The former view holds that the Earth was born violently, and knocked into shape painfully and traumatically by catastrophic upheavals and celestial bombardments. The latter view maintains that the crust of the Earth was formed by an excruciatingly slow process of evolution. Each terrestrial feature — the rocks, valleys and rivers — all changed at the same, slow, 'uniform' rate.

It is possible, of course, for the catastrophic perspective — still held by only a tiny minority of scientists — to regain some of the status it had in the 18th and 19th centuries. There is, for instance, now a growing consensus in favour of the idea that the dinosaurs *did* become extinct through catastrophic events.

In the meantime unorthodox theories about the Earth are more palatable to scientific opinion when framed within the unitarian school. This was done by no less eminent an astronomer than Thomas Gold. Writing in a 1955 *Nature* article he postulated the idea that the Earth may in fact have rolled over in slow motion several times in the past — a prolonged polar flip theory. If a continent the size of South America, surmised Gold, were to be raised by about 100 feet (say, through geotectonic movements) the axial spin and the axis of angular momentum would cause the planet to topple over at a rate of one degree per thousand years. In the process the location of the ice pack concentrations would be shifted towards the Equator.

Antarctica and the Weather Men

Antarctica, then, holds a fatal fascination for scientists. They now know it is the most treacherous, unstable, continent on Earth. Studying the Antarctic is still immensely rewarding, and not just for Doomsday theorists. It can yield vital geologic glaciological, climatic and atmospheric knowledge. Scholars from the US and Europe are pursuing seven separate but related projects right now.

But it is not without significance that the climatic perspective has dominated research in the last decade. Almost all of the world's climate moulding features — the oceans, land contours, the Sun — are sufficiently well known and understood. Yet no comprehensive picture of long-term weather patterns can be properly painted without an understanding of the global thermodynamic system as a whole. If Antarctica is left out of the picture our understanding of climatic change is incomplete.

Some glaciologists do believe the Earth may be in peril from Antarctica's instability, but not for the reasons the pole shift theorists give. The ice packs do not seem to be growing higher every year, threatening to topple the world over, but instead seem to be quietly melting. And this could pose a different kind of menace to the world's population.

Discoveries of what appeared to be a melting of Antarctica were made shortly after the Second World War. British press reports in October 1950 were speculating on the disappearance of islands in the Pacific Ocean after gales around Tierra del Fuego. Then it was believed that melting ice was partly the result of shifting weather zones, and an unusually warm gale that roared from the direction of the South Pacific.

A short while later a British scientific expedition went to Punta Arenas, Chile, to check on whether the sea level was rising, or instead whether the west coast of South America was, as alleged, collapsing into the sea. The British escort vessel 'Titan' reached the South Pole ice barrier, and cabled to Punta Arenas that the shipping lane was nearly free of ice floes and pack ice. It said 'enormous quantities' of Antarctic ice must have melted, adding that the 'temperature in (the) central ice district (is) four degrees higher than two years ago'.

Scientists at the time were divided on the significance of these findings, even though the magnitude of the warming was considerably higher than expected. Antarctica does, as one might expect, fluctuate seasonally. Like a thing alive the South Pole has the unique capability of greatly expanding and shrinking its own perimeters. The ice melts a little at each pole during the summer, and re-freezes new ice during the winter.

But Elmer Robinson, from Washington, is a meteorologist who, like others, is determined to check out the rumours of an Antarctic melting. He is concerned in particular with the impact of terrestrial pollution on the climate. How well, mused Robinson, does the polar ice, so crucial to the global heat balance and the weather machine, shape up under the atmospheric assault from the industrialized world?

Hitherto past technology has not been up to the task of finding out accurately enough. But, one by one, with the advent of sounding rockets, balloons, instrumented aircraft, and now computers and satellites, the barriers are falling. Robinson's team searched for a variety of compounds — nitrous oxides, methyl chloroform, methane and various carbon elements. Aerosol spray gas compounds that were detected at the South Pole in the 1970s were discovered to be partly absorbed by the ice. This was a new find, since it was thought that the gases, being relatively inert, would stay aloft virtually for ever.

Robinson now believes that the ice, which is fairly porous at the upper layers, may soak up fluorocarbons at low temperatures and retain them under accumulated snow. This is worrying, and confirms man's involvement in the present atmospheric warming. A study by Ralph Cicerone and colleagues at the National Centre for Atmospheric Research, published in May 1985, shows that some of the large Man-made molecules of trace gases, such as chlorofluorocarbons and cromotrifluoromethane, are far more effective in absorbing heat than CO^2.

Elmer Robinson was soon joined by other researchers, each equipped with ever more sophisticated testing equipment. A French scientific team is now, for the first time, using electrically heated drills. The drill melts the ice as it bores its way painstakingly down to the bedrock 9,100 feet below the ice. This is well over 2,000 feet deeper than the most up-to-date American drilling techniques will permit.

Drilling into the ice is still the best way of finding out what past climates were like. The continual sub-zero temperatures of the poles (ranging from minus 5C in summer to minus 40C in winter) are the result of the tiny amounts of solar radiation they absorb at such extreme latitudes, compared with the rest of the world. But there is often enough moisture in the air to cause snowfalls, and thus build up fresh ice. As each layer of snow is compressed under the crushing burden of newly deposited snow it becomes progressively denser. The air trapped between the grains of snow soon gets squeezed out, leaving minute air holes. The crystal structure starts to change, and stratified snow layers build up. This kind of compacted snow is known as *firn*, and under further pressure it becomes true glacier ice.

Scientists are capable of examining a compacted snowflake with its tiny fissured air pockets to find out the initial air temperature in which it was formed. In 1984 glaciologists refined a new technique for estimating the temperature of past climes by studying two isotopes of oxygen — oxygen-16 and 18 — both naturally present in air and water vapour and 'recorded' by the ice cores.

The ice can also show levels of past carbon dioxide concentrates. Working from 200-metre ice cores, Dr. B. Stauffer of the Physics Institute, Bern, Switzerland, has shown recently that CO_2 levels have risen from 280 ppm in 1750 to more than 345 ppm in 1984.

Slowly, then, the historic composition of past atmospheres, and their relationship with the advance and retreat of the Antarctic ice sheets becomes clearer.

The Unstable Ice Shelves

The Arctic, at the other end of the globe, is, in effect, a landlocked ocean basin. And its pack ice is floating on the ocean. However, as any whisky drinker knows, melting ice doesn't affect the level of the liquid in which it floats.

The Arctic is not then a present threat to the coastal areas of the world. But during a severe Ice Age the North Polar regions cause extensive glaciers to form in Northern Europe. One argument maintains that if the Arctic was not landlocked, the northern hemisphere would become glacier-bound much more often than it is. If the Arctic and sub-Arctic regions were open to the world's interconnecting oceans they would interact with the warmer oceans to the south, and hence be more subject to heavy snow showers. As it happens, northern Europe benefits from the warming Gulf Stream, but the Arctic — boxed in by land — does not.

Even so, the massive ice sheet left over from the last glaciation that now covers most of Greenland *is* vulnerable to a global warming. It extends for 700,000 square miles, and has an average thickness of a mile. If this ice sheet were to melt, the world's sea levels would rise about 30 feet.

Antarctica, on the other hand, is surrounded by the Antarctic

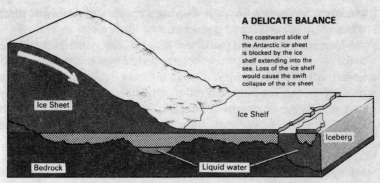

Source: Discover magazine (US), February, 1984

Ocean, the Atlantic, Pacific and Indian Oceans. If there were no ice in the South Polar lands explorers would be confronted with a clump of frozen rocky islands to the West, and a massive mountain ridge to the East.

The largest segment, East Antarctica, faces towards the Atlantic and Indian Oceans. It is a fairly typical stable land mass, consisting of a foundation of igneous and metamorphic rocks overlain by more recent sedimentary and volcanic rocks. West Antarctica, instead, is a series of mountain ranges whose rocks are much younger.

According to two Antarctic researchers, Joseph Fletcher and John Kelley, solar radiation at the South Pole is amplified by the extent of sea ice. Ice is a very sensitive lever that can magnify hugely the effects of even minute changes in the circulation of atmospheric heat. It in turn has a knock-on effect on the weather in the southern hemisphere. The North Pole, playing a less dominant role in air circulation patterns, follows the lead of the southern hemisphere — even the northern pack ice is determined by the southern climate.

Much scientific interest has lately focused on the so-called Ice Shelves, in particular the Ross Ice Shelf. This is an area of slab ice larger than France, and looks on the atlas like a wedge of ice that has been hammered inwards from the surrounding Ross Sea. What attracts attention is the curious and unusual configuration of the ice shelves. Is the ice resting on land, or is it floating? It seems to do both. The shelves in effect pin down the rest of the ice sheet, and prevent its breaking up and drifting apart. The shelves themselves are created by fast-flowing streams of ice. As the ice shelves build up, they 'calve', or break away, into icebergs. A steady state is generally maintained, whereby as much ice is lost by calving as is picked up from ice streams below the waves, and snowfall above them.

But the Ross Ice Shelf is inherently unstable, as it is only tentatively hinged onto rocky islands. Some glaciologists believe there is nothing to prevent it breaking up tomorrow, with or without the Greenhouse warming. Others think that only a massive heat-up, over a span of several hundred years, will put the Ross sheet at risk. So the controversy, and the scientific interest, is sustained.

Many scientists believe that Antarctica is not as thick as it was. Some think that the ice, under its own crushing weight, has flowed steadily outwards to form floating ice shelves. Others think otherwise. Anthony Gow, of the Cold Regions Research Laboratory, Hanover, New Jersey, believes that the edge of the ice sheet is either stationary or has actually retreated, because snow is simply stockpiling. But this itself is worrying, as it suggests a thaw in temperatures (remember that snow cannot fall at very low temperatures because the air will contain insufficient moisture).

Sion Shabtaie, originally from Iran but now a glaciologist at the University of Washington, believes water of unknown depth lies beneath the ice. The ice, then, is constantly — and dangerously — being lubricated by the water.

Herein lies the weakness of Mark Meier's theories. Meier is a leading glaciologist with the US Geological Survey in Tocama, Washington, and points to the build-up of Antarctic ice as a refutation of the Greenhouse Effect. But without a drastic return to Ice Age conditions, which would freeze solid the water lying below the ice and maintain a grip on the land, more snow and ice will inevitably continue to accumulate.

Moreover, Meier forgets about the shrinking albedo factor that would arise from a warming. Even a small rise in temperatures would have a much greater knock-on effect in the colder regions of the Earth as the snow packs and glaciers start to dissolve. Meier doesn't deny the Greenhouse Effect exists, merely that it is only about half the magnitude as predicted by the EPA and other institutions. Meier's is the worst of all worlds, with a slower-acting Greenhouse warming accompanied by more snow precipitation,

The Disintegration of Antarctica 109

the sooner to bring on a runaway slippage of the ice.

Glaciologists now believe that this slippage must sooner or later take place. Anthony Gow believes molecular friction will ultimately cause the lubricated ice sheets to slide forward. If it does this the entire region may crack up catastrophically, the slush acting as rollers. The dragging friction of the ground below would cause further melting. Gravity exacerbates the phenomenon, causing great masses of rock and mud to move downslope.

Such an event may have happened in the past. Some argue that it is difficult to suggest any other causal agent to account for the massive infusions of fresh water into the Gulf of Mexico. Cesare Emiliani's radiocarbon dating of tiny marine creatures called foraminifera shows that they died in the Gulf about 11,500 years ago. So great was this infusion that it raised the level of the world's oceans by 130 feet or so, flooding most of the coastal margins.

Scientists at the Australian National University also produced, in 1980, clear evidence from New Guinea coral reefs that the sea level suddenly became raised by 25 feet some 120,000 years ago. This was 'possibly' caused by a huge chunk of the Antarctic ice slipping into the sea. And some 95,000 years ago the sea rose even higher. John Hollin of the University of Colorado has mustered evidence from around the world to show that the seas rose 60 feet when millions of cubic miles of ice broke away from Antarctica.

The Hot Antarctic

We must remember that the Earth — including Antarctica — is permanently subject to the laws of physics that dictate that mass, energy and heat are interrelated properties. The Earth in fact has several heating systems independent of the Sun. The molten core is a product of the radioactive decay of matter, and the crust itself consists of constantly degrading radioactive minerals. The tectonic plates of the Earth are a source of friction heat as they continually collide with each other, even though the rate of movement involved may be no more than one inch a year.

In fact geological instability implies frictional heat. All of the circum-Pacific countries — Japan, Indonesia, the Americas — are prone to earthquakes as tectonic plates bump and grind together, melting granite and making the crust thinner and hotter.

This implies three things about Antarctica. Firstly, heat, radiating from the centre of the Earth, can melt the ice caps from the bottom upwards. Secondly, the possibility of an eruption beneath a glacier — where tremendous vulcanized heat meets with the extraordinary pressure of dense pack ice — is still a real one. The hot lava would liquify the ice for hundreds of feet over a very short time span. Devastating floods composed of icebergs as big as houses, plus volcanic debris, rocks and mud — not forgetting the millions of tons of meltwater — would ensue. Thirdly, as Richard Cameron of the American National Science Foundation recently argued, heat beneath Antarctica can be generated from the sheer pressure of mass increasing molecular action. As the ice cap slides outward from the centre of the polar region the friction would create further heat — a self-perpetuating effect.

In any event Antarctica today is not as cold as people think. As New Yorkers were shivering at −40 C in January 1985's freeze-up, at an Antarctic research site called McMurdo the temperature was a mere −3C! An extrapolated rise of 5C in the southern hemisphere would soon thaw out the ice. N.W. Young, of Australia's Dept of Science & Technology, points out the expected melt rate would be up to 10mm a day, 'sufficient to bring approximately the upper 1.5 metres of the snow pack to melting point'. The temperatures of the surrounding sea, says Young, would rise above freezing point (-1.8C).

In the meantime, warns Sion Shabtaie, if those ice shelves do give way, the entire Antarctic could dissolve in just 200 to 300 years. G.J. MacDonald, who has published a major study of the carbon dioxide effect on Earth, defers the breakup by another 200 years. He believes that the West Antarctic ice sheet could be set afloat within a 500 year time span after being triggered by a 5 to 10C warming. By themselves the Wilkes Land or the Filchner Ice shelf in the Weddell Sea could break up about a quarter of the total of grounded ice, an amount that had taken 7,000 years to grow. If this flopped into the sea it could cause tidal waves around the world.

Chapter Nine

THE FLOODWAVE EFFECT

It is curious to observe how Mankind seems to suffer from recurring apocalyptic nightmares. It is even more disturbing to note that occasionally such nightmares come true, and perhaps become stored in some kind of genetically transmitted folk memory.

When Man first learnt to write he recorded some of these traumatizing folk memories on papyrus as a warning for future generations. The Great Deluge, the Biblical flood of Noah, the primordial flood sagas of the Amerindians, the Aztecs, the Chinese and many other races, still retain a powerful hold on the human imagination. Indeed, we still use the term 'antideluvian' to describe history before that point. It was the doom-laden language of the Bible that first induced many early geologists to view Earth's history as entirely the product of catastrophic and violent forces.

Of course, many questions remain unanswered, including the precise historical epoch covered in the Genesis narrative and the exact nature of the 'deluge'. Even Christians are uncertain as to whether the Great Flood is historical fact or theological metaphor. Assuming it to be fact, then, the cause of the flood is still as obscure as the etymology of ancient words used to describe the event. The flood sagas seem largely Sumerian in origin, and Sumeria and Mesopotamia around 5000 BC suffered at the time from abnormally heavy rainfall. There were also many catastrophic river overflows in many parts of the world at that time (the Chinese legend hints at this by claiming that the over-spill of 'great rivers' was halted by the eventual swelling of the sea).

The evidence of the Earth itself, however, seems to give credence to the tidal wave explanation. The Floodwave attacks that destroyed many ancient communities were, in all likelihood, the end result of the rapid melting of the polar ice that was accumulated during the last great Ice Age.

Indeed, much of the worry about a rapid melting of the world's ice arises from what is reckoned to have happened at the end of the last glaciation. The advances in geophysics — in particular paleoclimatology — have greatly aided our knowledge of past Earthly environments. We know that the cause of the warming last time was different. The population of the human race was far too insignificant to affect the climate, unlike the situation obtaining at present. Bill Kellogg, of the National Centre for Atmospheric Research, Colorado, instead believes that the warming of 4,000 to 8,000 years ago during the Altithermal (Climatic Optimum) was probably due to a change in solar radiation, or had something to do with the Milankovitch Effect.

Certainly there is much evidence to show that the change in temperatures was rapid. Undoubtedly it took, by anthropological standards, a long time — from 10 to 15,000 years. But this is an extremely short episode in geologic time — and it was also very recent.

Indeed, the Ice Age, usually given in the literature as ending between 12 and 10,000 years ago, probably ended later than is conventionally supposed. Glaciologist John Mercer, of Ohio State University, says that the critical component in the melting, the Ross Ice Shelf, shrunk backwards from the mouth of the Ross Sea to its present position between 14,000 and 7,000 years ago. This breakup of Ross was entirely, he says, due to the unnaturally warm temperature of the oceans. It was this, rather than any *atmospheric* warming, that caused the ice to shrink, slip, and finally disintegrate.

We must remember that at the time of the start of the last glaciation, between 500,000 and 100,000 years ago, there would have been as much as 18 million cubic miles of ice; more than twice the amount existing today. The unified fronts of the ice packs would have covered more than 30% of the Earth's surface. In the northern hemisphere, known as the Laurentide, and mainly concentrated in what is now Greenland, this ice would have covered present-day North America down as far as Oklahoma, the whole of Scandinavia and virtually all of Europe. The water withdrawn from the oceans to feed the glaciers and Antarctica would, at the height of the Ice Age, have been some 12% of the total, thus dropping the ocean levels by a maximum of 300 feet or so.

Hence, if a warming of the rapidity and magnitude envisaged by Mercer actually took place, the ice would have dispersed at an alarming speed. In some parts of the world, and at certain times, the sea level would have been rising as fast as *20 feet per century*. There would have been a noticeable increase in wetness along coastal areas during the lifetime of one long-lived adult. There would have been extensive flooding, with sea defences being destroyed and village communities forced to flee further inland.

This rate of tidal increase, mercifully, would not have been constant. Stephen Schneider suggests that sea levels probably rose by about half an inch a year for long periods. But this was interspersed with much faster rates of rise, exacerbated by massive coastal land slips.

Much of this fluctuation in the speed of sea level rise can be explained by the prevailing temperature of the oceans. Warming will increase the density of ocean water, and thus its volume. In other words the water will take up more space. An added complication is that the thermal expansion of the oceans will affect Antarctica as well, as John Mercer pointed out. Any rise in sea level will bring the Ross Ice Shelf breakup that much nearer.

Furthermore, a rising sea level will itself bring about an additional warming because of the lowering of the altitude of continents. And, as anyone who has paddled in the warm beach pools left behind by the retreating tide will know, it is clear that the new, shallower, heat-absorbing coastal seas and freshly created bays will warm up the atmosphere still further. The climate, in effect, will struggle to become more uniform, as the increased area under warmer water will permit a greater transfer of heat from the equator to the poles.

But it is not just rising sea levels that would have caused problems for the inhabitants of a post-glacial civilization. Something else happens when massive quantities of ice dissolve. In the first place the crushing weight of glaciers — over tens of thousands of years — must have compressed the crust of the Earth to below its pre-Ice Age elevations. And what goes down must, eventually, go up. If the ice covering lost more than 50 centimetres in height per year this would amount to 15 million metric megatons of weight loss within a decade.

Hence the kind of ice loss we are talking about would have engendered colossal landslips and volcanic eruptions, as massive crustal upheavals took place. Some scientists believe that a series of violent earthquakes accompanied the end of the Ice Age, possibly accounting for much of the tumultuous language in which the flood legends were couched. These vivid and dramatic events would have sent literal shock waves through all the world's interconnecting oceans, and would have been recorded for posterity by the learned elders of ancient civilizations.

Indeed, it is interesting to note that the Egyptian account of the flood does not actually use the word 'rain'. Some theologians no doubt secretly prefer the Egyptian to the Hebrew version of Genesis, since not even the most geographically illiterate layman could believe that a mere 40 days and 40 nights of rain would drown the entire world (although it might just fill a dried-up Welsh reservoir).

The Egyptian legend, reporting that water 'fell' on the land, is revealing. This key word immediately suggests a tidal wave, possibly a *tsunami*. Amongst the tribes who lived in South America at the time in places like present-day Peru and Chile, seldom is any other type of flood experienced.

Tsunami is a Japanese word, and originally meant any form of tidal wave (Tsu = harbour; Nami = wave), but in this century it has come to mean a sea quake or seismic wave. They are caused by tectonic disturbances beneath the ocean floor that generate waves travelling outward at phenomenal speeds. They often reach enormous heights when they arrive in shallow lagoons near the land, frequently taking vacationers by surprise. They can also be caused by volcanic eruptions, submarine landslides, and even by nearby man-made explosions.

Two of the world's worst *tsunami* disasters occurred within a few years of each other. One struck the shore of the Bay of Bengal in 1876, and left nearly 200,000 dead. The other was the famous volcano that blew up in 1883 in Krakatoa, East Indies. It has preoccupied the attention of climatologists like Hubert Lamb because of the millions of tons of climate-moulding ash that was belched into the skies. And yet the real harm that Krakatoa caused was when huge tidal waves flooded the coasts of densely populated neighbouring islands. The shock waves from Krakatoa even affected the English Channel, on the other side of the globe.

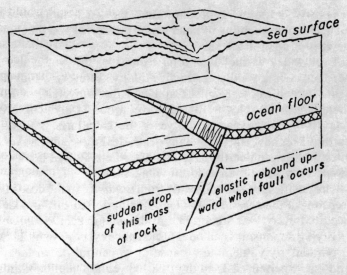

A tsunami is caused when a fault in the crustal rock causes the sea floor to drop rapidly. The sea surface above also falls, causing massive seismic sea waves to race away at high speed.

It is clear that when massive seismic disturbances take place close to, or beneath, seas and oceans it is not just underground rumblings that cause tidal waves. These would more likely cause a *seiche*, which are rhythmic oscillations of water that often affect an enclosed body of water such as a lake or bay, giving rise to waves no more than five feet high.

Vulcanism or earthquakes can cause, instead, a violent vertical displacement of the seabed itself. Since water is not compressible, an entire column of water, from floor to surface, is set in motion to race away from the seismic zone. In the open ocean the waves are no more than a few inches high, even though travelling at high speed. When they pass a ship on the high seas, those on board probably don't even notice them. And this in spite of the waves travelling at their fastest. The deeper the water the faster the waves travel as the speed of the wave is equal to the square root of the product of the acceleration and the depth of water.

A *tsunami* may reach a mere 30 mph in 60 feet of water, but exceed 600 mph in 30,000 feet of water. The danger arises when the *tsunami* is on its last lap, heading towards the narrow coastal reaches where its energy becomes concentrated. It then towers into an ugly looking column of water which then breaks on the shore at heights of up to 100 feet.

Even today earthquake-prone regions of the world, like the Mediterranean, yield a destructive tidal wave every two or three years. According to Michel Caputo of the University of Rome there have been at least 110 tidal waves around Italy's coast since AD 1000. In the past three years, he calculates, land-based earthquakes have triggered many death-dealing seismic waves. He cites the Messina earthquake of 1908, where 60-foot tidal waves swept away more than a dozen villages.

The Pacific area suffers greatly from *tsunamis*. In the deep ocean trenches of the ocean itself at least one undersea earthquake has occurred every year since 1800. And a major Pacific *tsunami* of terrifying dimensions can be expected once every 10 years. Hawaii is particularly vulnerable because of its central ocean location, and has experienced 37 in the past 130 years. The inhabitants of Californian coast cities have suffered from seismic sea waves originating several hundreds or thousands of miles away. In 1964 about $10 million worth of damage was caused to the Californian coast when an Alaskan earthquake shook America. In Crescent City alone, where 30 blocks were flooded, $7 million worth of damage was wreaked.

The Menace of the Tides

The suggestion that the world has only recently come out of an Ice Age is based on the observation that much of the Earth is still rising (i.e. the land is still rebounding). For example, as we shall see, the south-east of Britain is said to be sinking at a rate of a foot per century, while the north of Scotland is rising, as it recovers from the weight of the ice.

But this fact poses an intriguing anomaly. If the Earth has not fully recovered from the last Ice Age, how can there still be a threat of a further melting of Antarctica?

The answer was spelled out in the last chapter. There is still an awful lot of pack ice around in the polar regions. This ice may in fact represent the optimum, or median, amount of ice that has always existed during *interglacial* periods (like the present). Probably over half of the 18 million cubic miles of Pleistocene ice melted, but by no means all of it. Now, however, the human race is in the unparalleled and extraordinary position of being able to melt nearly *all* of the present ice sheets. Sea levels will rise catastrophically as a result, and lowland dwellers will be forced to move further inland.

Clearly, much coastal flooding will be inevitable with or without climatic change. Earth is a very watery planet. The total surface area covered by all the world's oceans is nearly 140 million square miles, or some 71% of the Earth's surface. The quantity of oceanic water represents an enormous 325 million cubic miles. If it were accumulated into one place it would form a sphere about 864 miles in diameter — probably larger than all the asteroids or planetisimals in the Solar System put together. If the oceans were divided up each man, woman and child would get 11,000 *million* gallons.

If the lithosphere — the solid ball of the Earth itself — was completely spherical the entire planet would be uniformly covered in water to a depth of almost 8,000 feet. The tidal currents would be hardly noticeable. But the sizeable 30% of the dry land surfaces of the world succeed in breaking the paths of these currents, often with highly damaging results. The Pacific tides continually crash into Asia and Australia, while the Atlantic tides collide with America. Certain topographical features, moreover, ensure that the Floodwave Effect is more severe than it would be elsewhere. The Gulf of Mexico, for example, has an awkward funnel shaped by the adjoining coasts of Louisiana and Mississippi. Around the world narrowing, constricting coastlines and shallow estuaries sandwich the water so that its height builds up.

The tidal wave soon becomes a *tidal bore*. These have reached 15 feet in height in the Amazon, and even greater heights have been described for the Tsientse Kiang in northern China — occasionally as high as 25 feet. Britain in particular suffers from having a number of narrow funnels which can produce tides rising to 40 feet, such as the Bristol Channel and Severn Estuary, up which frequently rolls the Severn Bore.

Unceasing geotectonic pressures mean that the land is doomed to continually change places with the sea. But this makes predicting future sea level changes rather difficult. The changing shape of the ocean floors also shifts their contours over time. Some earth movements might add to — and others counterbalance — the rising tide. Sudden land slippages can abruptly turn a steady two-foot-per-century rate of sea encroachment into a 10-foot-per-century rate.

In the past the world's shorelines have been maintained by the balance between sediment gained and lost, between the sludge dragged out to sea in the middle of storms, and that dumped back onto the shoreline by the whiplash tides. The danger from rising tides is that they enable storm waves to erode a beach further inland by a much greater amount. The higher the tide the greater 'reach' it has. For instance, a rise in sea level of just one foot would erode most sandy beaches by at least 100 feet.

Cities Under Seige
A glance at the atlas shows that, with very few exceptions, the largest cities of the world lie near or on the coast. All great civilizations grew up along riversides, or near important waterways or ports. Paris, Washington and Bonn, for example, all lie on a series of high river terraces. The three largest and most densely populated megacities — New York, London and Tokyo — are coastal seaports, as are Singapore, Columbo, Hong Kong, Sydney, San Francisco and Los Angeles.

Nowadays an estimated half a billion people live on floodplains, with untold millions more living along vulnerable coastal areas subject to tidal waves. Perhaps as many as 200 million Chinese live in the Hwango Ho and Yangtse River valleys.

Mankind is therefore used to, or at least has tolerated, the watery sector of the Earth. It has provided immense benefits to growing civilizations, enabling fishing and mercantile communities to thrive, with rivers aiding transport, and dried salt water helping to keep meat products fresh. Those living in coastal regions, of course, had the most to fear. The waves and currents relentlessly attacked the land, eroding here and depositing there, with the occasional tidal wave demolishing whole fishing communities that had become established too near the coast.

For primitive communities by the sea the expense of coastal defence at periods of rapid aquatic invasions would have been astronomical and probably futile in the long run. Barriers even 150 feet high would sooner or later have collapsed, and presented an unacceptable risk to those living by the increasingly turbulent sea.

Even today there have always been cities, often historic and

The Floodwave Effect

prestigious, that have suffered unjustly from the rampaging waves. Let us take Venice, jammed between the coasts of Yugoslavia and eastern Italy. The high tides that have washed into the lagoons and canals that surround this tourist mecca have become increasingly grave.

Venice is right now sinking into the lagoon, and most piazzas and parlours lie no higher than 30 inches above sea level, with the canals lapping against crumbling stuccoe walls. Since 1876, the year when detailed record keeping commenced, high water generally occurred twice a year at most. By 1930 that average had become seven times a year. In the 1950s the figure had increased to 16 times a year; today it is twice that. The most serious flood occurred in November 1966 — the month when serious river flooding devastated Florence and ruined millions of pounds worth of historic works of art.

Venice's problems were compounded long ago by the blocking up of a number of artesian wells by the local authorities. The fate of the city was still determined by a sadly neglected system of dikes more than 400 years old which linked the islands separating the lagoon from the Adriatic. So lower portions of the town, like St. Mark's Square, are often inundated. But the tourists see little of the drama, as the high water threat comes mainly in November and December. Then fierce winds gust in from North Africa and create a mini surge, causing huge waves to batter a prominent reef that separates the lagoon of Venice from the sea.

Bangkok, Thailand, is also subsiding for similar man-made reasons. As the population of the city expanded following the Second World War, this 200-year-old trading village — set in marshy lowlands on the banks of the Chao Phraya River — began to suffer a water shortage. So private industries, hotel and housing estates started to sink their own artesian wells deep into Bangkok's water table.

When the quality of water began to deteriorate, the wells were drilled deeper. Eventually the underground reservoir ran dry, and the city began to sink at the alarming rate of about six inches a year. One shopping mall needed new steps to bridge a massive cavity at the base of the wall, and cracks in office buildings appeared. Parts of the capital have dropped by as much as three feet within 15 years. Today Bangkok is fast living up to its name of 'the Venice of the East'. Prinya Nutalaya, of the Asian Institute of

Technology, now fears that as the land is only about three feet above sea level the entire city will be under water by 2000 AD.

Big modern cities are prone to flooding for other man-made reasons. Tracts of porous soil are covered over by impervious surfaces, and flat, hard roadways greatly facilitate the thousands of gallons of murky water from overspilling rivers or canals to penetrate vital residential and business areas.

Bangkok again provides a prime example of this. As the city prospered and many citizens bought cars and took to the cramped roads, the town planners filled in the canals and covered them with tarmacadam. The city soon began to experience severe flooding every year, as the natural monsoon flood control mechanism had been destroyed. From June to October up to 45 inches of rain can fall in Thailand, with the canals acting as a stormwater release mechanism. Even in a bad flood, it was always easier to get out of the city in flat-bottomed boats. Now the city frequently comes to a standstill as the streets rapidly become waterlogged.

What finally clinched the deal to build the mammoth Thames Barrier at Woolwich, South London, was alarming evidence that, since the 15th century, the south-east of England has been sinking by about one foot per hundred years. London, built on a bed of clay, is subsiding even faster than the surrounding country.

Furthermore, the sinking has not always been gradual. During the last 2,000 years the south-east has twice subsided, by all accounts quite suddenly. Massive inundation in the Thames Valley occurred each time. And it was the second subsidence of the land surface at the beginning of the 11th century that greatly worsened the flood risk in Britain. The gradients of the river channels eased out, and the rate of flow slackened. There was a dramatic increase in pressure on sea defences and river walls.

Indeed, the evidence of Roman sites has confirmed that the land surface is now 15 feet lower than it was during the early days of the occupation. For example, human artefacts, and the bones of land animals, have been dredged from the bottom of the North Sea. In 1878 Roman articles were uncovered in the Essex tidal mud, proving that habitable land extended further out to the east than at present. At West Tilbury a Romano-British hut circle was found buried on the Thames foreshore about 13 feet below high water mark at Clacton. Either coupled with subsidence, or for other reasons, sea heights increased. Dutch engineers and dike builders believe that a rise in sea level was not slow and progressive, but occurred in short and destructive bursts at about 200 year intervals.

Hence thousands of acres of Lincolnshire were flooded by the tide in AD 245. And in AD 419 the Hampshire coast was drowned. Old Dunwich on the Suffolk coast — a prosperous merchant port in medieval times — was pulverised out of existence by the storm-lashed waves. By this time the whole of Northern Europe was in chaos. At the time Dunwich was being pounded, sea floods on the Continent were drowning people in their thousands. Some 400,000 alone were reported to have died in a 1570 disaster. Ancient annals recorded that in 1634 there were 'great losses of land from the Danish and German coasts'.

Owing to the sharing of the English Channel coastlines, France has also suffered from disastrous sea encroachment in postglacial times. Britain became separated from the Continent within the last 65 million years, when the Thames ceased to be a tributary of the Rhine. Seafloor spreading was made worse by vertical slippage as the swelling Atlantic bulldozed its way through the Straights of Dover towards Norway. France is even now said to be sinking by about 30 centimetres a century.

The Dutch are no stranger to floods — hence their expertise in dike building and land drainage. Dutch sea walls collapsed in December 1287, and again in November 1421 the sea burst through dikes, flooding 72 villages and drowning an estimated 10,000 people.

The unstable geology of north-west Europe helps explain why London, more than other European cities, is particularly vulnerable to flooding. This massive, rambling metropolis is built around the Thames, which is more of a tidal estuary than a river. Estuaries are particularly vulnerable to the 'surge' (rather similar to a bore), where a wild sea, swollen by low atmospheric pressure, becomes jammed between two narrowing land masses, and so surges up the nearest watery opening. The mouth of the Thames is just such an opening, and the long and chequered history of Thames tidal overspills has been spelt out in my book *London's Drowning* (1982).

To make matters worse, the record shows that — from 1820 to the present — the sea has been rising with ever-increasing frequency.

In addition coastal storms appear to occur in long cycles which themselves seem to be decreasing in their time span. Since 1845 the gap between combinations of tide and surge gaining on previous maximum levels has grown smaller. For example, since the great London floods of 1953 — which have now become a celebrated yardstick with which to measure flooding levels — the maximum height has been reached three times. In January 1976 an 8½ foot surge caused widespread flooding in the estuary areas.

The fact that the flooding threat is taken seriously by the authorities is evidenced by the millions of pounds spent in recent years to keep the ever-swelling tide permanently out of the capital. Some 30 years after the 1953 floods — in which 307 died — the massive and unique rising-sector Thames Barrier was officially opened, and is now a visible reminder of the threat to Londoners from a combination of high tides, sinking land levels and progressively rising seas.

Future Floodwave Scenarios

Taking a middle-range assumption as being the most likely scenario for the future (and discounting an alternative low and high scenario as being too unrealistic), from now until the end of the century, according to the EPA, 'sea level will rise almost as much as it has in the last century'. Even using EPA's most optimistic model, sea level around the world in the next 20 years will rise twice as fast as its normal historical average. And from 2000 AD to 2025 AD onwards the tides will rise three times faster than the historic rate.

The National Academy of Sciences foresees a two-foot rise in tidal levels by the end of the 21st century. The EPA, however, because of their reliance on different models which predict the potential impact of trace gases, is much more pessimistic about the longer term. The Academy's two-foot rise will be achieved long before 2099 AD. Indeed, by 2040 the seas will definitely be 2½ feet higher than they are now, with trends accelerating as time passes. They consider a global rise of between 4.8 feet and seven feet as 'most likely'. And a rise 'as high as 11 feet by 2100 cannot be ruled out'.

And in two to five centuries from now, say Basil Booth and Frank Fitch, the Netherlands, the Po Valley in northern Italy, Northern Germany, Denmark and the northern shores of the

Soviet Union 'would all sink below the advancing tide of rising waters'. Writing in their seminal geophysical study, *Earthshock* (1979), they say that large parts of the Middle East, Bangladesh (long the victim of gigantic hurricane floodwaves), China and South America 'would be particularly badly hit'.

There is now unmistakeable geophysical evidence of quite recent changes in the shape of America's coastlines. Large segments of the northeast and northwest coast of the United States are referred to by geographers as 'drowned topography', meaning that the land has sunk relative to sea level. This, they say, accounts for the humpy, irregular nature of the shoreline from the Hudson River Valley to Maine, where the hills and valleys of earlier epochs have now been swallowed by the waves.

On the other hand most of the rest of the east coast from New Jersey to Florida is flat and straight, because the submerged terrain has a long gentle slope of continental shelf extending outwards for hundreds of miles. America's eastern and Gulf coasts are protected by an uneven picket chain of scrawny, sandy islands. These are known as barrier islands, and extend from Fire Island, New York, to Padre Island, Texas, protecting the coast from the excesses of soil erosion and flooding. The problem is that many of them are large enough to become tourist playgrounds. Cape Cod, Miami Beach and Hilton Head in South Carolina are examples. But the barrier islands are seldom permanent, and can vanish completely during a violent storm. And in recent years such storms have been particularly virulent.

In the meantime on the Alaskan coast a number of bays that were used as harbours a century ago are now too shallow to be navigated by ships. Much of the central Californian coast, from Monterey to Mendocino, is rising. And the west coast of America is slowly uplifting and tilting towards the east. The east coast, south of New York, is also gradually subsiding. North of the 'Big Apple', however, the land is still rebounding into its original ice-free contours.

In 1971 the US Corps of Engineers estimated it would cost $1.8 billion at 1971 prices to protect the most critical coastal areas, with an annual $73 million in maintenance costs. The Corps claimed that Louisiana was losing coastal land at a rate of 42.7 square kilometres a year. Some 85 square kilometres of land along the Virginian shoreline around Chesapeake Bay has been eroded

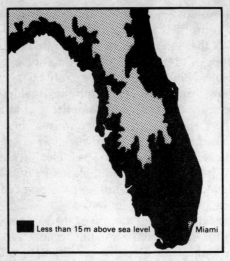

Within 200 years the coastline of Florida will be greatly reduced, with Miami submerged.

Source: New Scientist 12/6/86

during the period 1850–1950. Cliffs near Chicago lost 9.3 metres of rocks and chalk in the 11 year period from 1964–75. New York at the Battery is now barely four feet above the Atlantic coast at high tide.

Jim Titus, an economist who helped prepare the EPA report on future sea level rise, reckons that beaches have been eroded some six inches since 1890 as a result of the relentlessly rising sea. According to reports from the US Coast and Geodetic Survey the margins of the Atlantic coast have risen about half a foot in the last 80 years, with the greater part of that rise coming since the war.

Some parts of America, therefore, are already at a disadvantage even before the melting begins to seriously take its toll. Jim Titus reckons that US cities already below sea level, like New Orleans, would be the first to experience serious flooding. Whereas Venice is in danger of sinking into the sea, New Orleans has always been about eight feet below it, thanks to the French who built the city in such soggy surroundings. The city causes $19 million worth of problems every year just to keep it dry and upright. For a century or more the risk of building higher than four storeys was considered

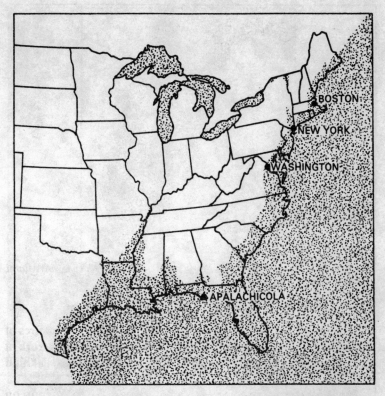

Likely extent of coastal flooding in the eastern United States in the event of a 250 foot rise in sea levels.

too great. Even now the tall blocks of downtown New Orleans are built atop enormously deep concrete piers. Charleston is in a similar position. A further five foot rise would inundate a quarter of the city. Charleston, instead of being flooded once every 100 years, as at present, would be awash every 10 years.

Some experts, by extrapolating present warming trends into future centuries, paint an alarming picture. By the end of this century it is reckoned that the Thames Barrier might need to be raised several times a year. Then, in perhaps 50 years the surge tides will rise as high as the top of those great helmet-shaped piers between which the semi-circular underwater gates are hitched. Then, as time passes, the surge tide will eventually break through.

The extent of European flooding in the event of sea levels rising 200 feet. The shaded areas indicate the height of terrain above sea level.

And then what? The gloomiest predictions suggest that the new shore line could reach to Piccadilly. New Scotland Yard would become inaccessible. London's 12,000 electricity sub-stations would be shorted out and require weeks, if not months, of drying out and repair. The Tower of London would be flooded out, and both Buckingham Palace and Parliament would be under feet of water. Booth and Fitch paint the grimmest of all scenarios for the capital. If, they surmise, the melting of the world's ice packs occurred over 200 years, London would become inundated during the first six. And even if the ice took 1,000 years to melt, 'London would flood in 28 years, perhaps less'.

Across America there will be widespread dislocation to civic life and enormous inconvenience for those forced to relocate homes and businesses. Stephen Schneider, working in collaboration with fellow climatologist Robert Chen at the National Centre for Atmospheric Research (NCAR), believes a rise in sea level of 15 to 28 feet would affect nearly 6% of the American population directly. A 15-foot rise in sea level would flood over a quarter of Florida. By the middle of next century, and at the latest by 2100, the entire southern tip of Florida, including all the Miami/Fort Lauderdale megalopolis, will be under water. Every city along the coast — Fort Myers, Tampa, Daytona Beach, will be flooded. Later, from the air, only the tips of the 'white refrigerator' beachfront motels of Miami Beach will mark the southern part of the state's location.

And according to Stephen Leatherman, in a recent research paper focusing on Galveston, Texas, a rise of five feet would double the already vulnerable Bay area subject to flooding every 15 years. Houston and Galveston already lie at or near sea level. In addition they both pull immense quantities of water from deep beneath the ground to feed the demands of industry. They are both in a Bangkok-type situation, with land surfaces sinking in wellhole cavities. Over the past 50 years, some 1,300 square miles of coastal land has subsided by more than a foot. And as the climate deteriorated the cost of storm damage in the Galveston area would jump from $23 million to $105 million. Sea flooding would also claim much of the Mississippi Valley. States like Louisiana would be awash, as would most of Alabama and large areas of Arkansas and Tennessee.

The flooding would not be unremitting, but would come in short, disruptive bursts, every few years or so. There would be plenty of time for mopping up until the next onslaught. But we should not forget that the aftermath would be as crippling as the flooding. Complete recovery would take a long time, partly because of a troublesome feature known as 'ponding'. This is floodwater which fills up in the low-lying areas — flat districts with poor drainage — and is unable to flow back quickly when the water recedes, and stagnates. The ground that had suffered from ponding would also become silted up, and would lose many of its life-giving minerals. Parks and gardens in the flooded zone would, for something like 40 years or more, never again be able to grow the same quality of trees and plants that they had before.

For the immediate future, however, up to the year 2100, the general consensus is that it is unlikely sea levels will rise higher than about 20 feet. Given our present state of knowledge precise details are still not forthcoming. Hence present estimates seem to outsiders to resemble an auction, with proferred tidal levels based largely on the subjective disposition of the researchers concerned.

Some scientists demand much more information about ocean currents, the salinity of the water around the polar regions, air temperatures, and the true role of the clouds (still subject to much speculation) before they will rush to judgement. A polar workshop sponsored by America's National Research Council (NRC) in late 1984 concluded, after reviewing the evidence, that sea level rise by 2100 would probably be no more than 18 inches. Roger Revelle, the oceanographer, by adding in the often neglected effects of the thermal expansion of seawater, raised the figure to at least two feet.

This is still a serious enough threat to the world's coastlines. But, as we shall see, rising tidal waves are only half of the problem. With a warming on the scale indicated the world's weather patterns will also be drastically altered.

Chapter Ten:

FLOODSHOCK!

There can be little doubt that the warming is making the world a more watery place. Floods are already in the big league of disasters, since they claim far more lives each time they strike. Only droughts affect more people. But floods are fast heading for the Number One position. For example the number of victims affected by floods rocketed from 5.2 million a year in the 1960s to 15.4 million in the 1970s. This phenomenal 300% leap, documented by the US Office for Foreign Disasters Assistance, is much higher than that for droughts, which only experienced a 35% increase.

Over 1964-82 floods affected the lives of 221 million people across the globe. Let us look at a few examples of natural disasters occurring in the 1970s:

Location	Type	Date	Estimated deaths
Bangladesh	Flood	1970	500–800,000
China	Earthquake	1976	242,000
Ethiopia	Drought	1973/4	200,000
Sahel	Drought	1970/3	100–150,000

The increase in the number of floods is the end-result of all the environmental disbenefits that have been documented in this book: over-population, deforestation, soil erosion, thermal pollution, the growth of megacities — and the resultant warming of the atmosphere coupled with the simultaneous heat-up and expansion of the oceans.

Climatologists have built up a pretty good picture of what the pattern of global rainfall was like during the Altithermal period of

4,000 to 8,000 years ago, when temperatures were fairly similar to what they will shortly be approaching. Using this as a yardstick, both the EPA and the NAS now believe that the warming will play havoc with the world's rainfall patterns and storm tracks. Many places will become much wetter, and others drier, than average. There will be climatic instability and extremes of seasonal temperature as the world's atmospheric circulation patterns change.

Temperatures will rise in the higher latitudes, and in the continental interiors of America, Europe and Russia. Although aridity will marginally increase and rainfall become slightly less regular in these regions, hurricane type storms will grow in frequency and intensity. As one moves further south air temperatures will fall slightly — computer models show that Japan, India, Asia and the Middle East will be that tiny bit cooler and wetter. These areas, though, will also probably experience far more violent storms than they are presently used to.

The reason? All the ingredients that cause rain and wind — atmospheric moisture and heat — are at the moment on the increase. What causes precipitation is a clash of temperatures as a warm front of air rises to meet a colder front, which cools the moisture in the air, which then forms into droplets of rain.

However, there are three things that can create *more* precipitation than we get at the moment: more living matter, both animal and vegetation (although at present the former, in the form of humans, is gaining at the expense of the latter), more heat (either terrestrial or solar), and more and larger cities, with their heat-island effect and the extra precipitation they cause (see Chapter Two).

This doesn't mean that the *total* quantity of liquid on the Earth will increase. The hydrosphere is a closed system, after all; whatever is lost in transevaporation is ultimately returned to the lakes and oceans of the world. But if there is more evaporation, rainfall will either be more frequent or more intense.

One of the most important storm-inducing features of the Earth is warm ocean water. Meteorologists are well aware of the havoc that this phenomenon can cause to the finely tuned weather machine. Recently the world's media paid great attention to the US National Oceanic and Atmospheric Administration's discovery that the Pacific Ocean had suddenly warmed up by some 11F.

The Pacific, being the world's largest ocean, is a major

component in the global heat engine. In July 1983 a great swathe of swollen warm water was discovered surging its way to the coast of South America. The weathermen borrowed a name often used by South American fishermen to explain this phenomenon — 'El Nino'.

It was a prime example of how relaxed air pressure interacting with warm sea currents could have far-reaching climatic effects. Surface water in the Pacific that would normally be pushing anticlockwise round the so-called Peru Current was racing back instead towards North and South America. Accompanied by strong air currents and torrential rain, it set off a chain of unpredictable disturbances reaching far into the jet-stream currents snaking 40,000 feet above the Earth.

Climatologists believe it was El Nino that was to blame for the schizoid character of the weather during 1983/4. Throughout the southern hemisphere many countries were suffering from floods and droughts at the same time, in different regions. The patchwork schemata was particularly marked in Thailand, where one village suffered drought while its neighbour barely 30 miles distant was inundated. Most of the floods were devastating. Parched scrubland had been turned into swamps, and torrential rains pounded Paraguay and Uruguay.

From 1980 Brazil had suffered from its worst drought this century which led to the deaths of three million. The drought ended abruptly in the spring of 1985 when floods caused many hundreds of deaths, and made nearly one million homeless. Ironically much of the flood damage was caused when more than 1,000 newly built reservoirs (to help cope with the drought) overflowed. What happened in Brazil in 1985 was merely a repeat of what happened to Peru and Bolivia two years earlier — drought followed by floods.

The outlook for individual countries looks stormy. In March 1986 crippling floods made 300,000 homeless in Bolivia as Lake Titicaca, near the border with Peru, overflowed. Peru's port city of Puno was inundated, with homes destroyed by mudslides. In July more than 100 people had been killed and hundreds made homeless in the Far East, including South China and Japan, and the Philippines. In August continual and torrential rain in Jamaica left 50 dead, having killed some 33 in Cuba and Haiti.

Already the ocean warming seems to be instigating more

monsoon floods in India. As the predicted new summer low pressure-centres originating over the northern plains draw in the still warm air from both the Indian Ocean and the western Pacific, hurricanes and storms are increasing. Every year some 700 people now lose their lives in Indian floods, as do 40,000 cattle, with the damage cost amounting to £100 million. The annual devastation caused by the *480 inches* of rain that falls over the Khasi Hills in Assam might actually be improved on. In August 1986 three million fled flooding in India as the Godavari River in the southeastern regions overflowed following two years of drought in Andhra Pradesh. More than 500 villages were submerged. One month later torrential rains caused widespread flooding in Calcutta and West Bengal, leaving nearly threequarters of a million homeless. Flooding in India's eastern Bihar state took the lives of some 300 people,.and by October 1986 42 lives had been lost in Bangladesh, with 100,000 left homeless. Overflowing rivers had left more than 100,000 people homeless and 1,500 square kilometres of crops ruined.

If, as the climatologists predict, America will get marginally warmer weather she too may well get more inland floods. This can be explained by comparing rainfall levels of countries situated at different latitudes. Boston receives nearly twice as much rain annually as London because her July temperatures are about 8F warmer. Hence the city loses much more summer moisture to evaporation, and is therefore more prone to cloudbursts. In the short term, before the threat of aridity becomes real, the mixed urban and agricultural topography of America will prevent the climate becoming too dry. It will instead become more humid, thus intensifying the power of tornadoes and hurricanes that pummel the Pacific and southern Atlantic seaboard, while giant tides continue to rip into the mainland.

The Drowning of America
In fact, the warming is already adversely affecting America's weather, as we have seen. The winter of 1982/3 was marked by astonishing extremes of climate that saturated 350,000 acres of Californian farmland. In some parts of the western mountain regions more than 600 inches of snow fell, to cause devastating floods when it melted. The following spring was an onslaught of tornadoes, thunderstorms and blizzards. In February 1986 killer

storms pounded California, leaving 10 dead and driving 10,000 from their homes.

There is today a growing threat from America's over-full rivers. According to James Cornell in his *International Disaster Book* (1976), overspilling rivers represents the most widespread geophysical hazard in the US, accounting for more annual property damage than any other type of disaster. The population living on lands virtually borrowed from rivers or the sea is over twice that of the national population density average. There are now about 50 million acres of US land known to be below flood level, mainly the most fertile, the most densely occupied and the most economically active.

River overflows are of major consequence for some 10 million people, and indirectly interfere with the lives of another 25 million. Cornell believes that American floods cost 60 lives and $1 billion in property damage every single year, compared with an inflation adjusted loss of $1 million at the turn of the century.

The dislocation that would be entailed today in a major US inland flood would be colossal. Even in July 1951 about half a million people had to evacuate their homes after the Kansas and lower Missouri rivers overflowed. Three times that number were directly affected in January 1937 when the River Ohio rose 20 feet above the expected high after widespread and persistent rain.

Now the Army Corps of Engineers is desperately trying to prevent the Mississippi River from running amok as each year it seems to get broader and deeper, and collect yet more silt as it lumbers downhill across the chest of America from Minnesota to the Gulf of Mexico. It drains the run-off from 41 states (and parts of Canada, as well) emanating from a number of troublesome, overspilling tributaries, such as the Ohio, Arkansas, the Red, the Illinois and the Des Moines. At St. Louis, where the Mississippi and the Missouri meet, the river is one mile wide. The Indians, in awe, named it The Father of Waters.

But as the weather turns increasingly neurotic this giant river is threatening to take a shorter and steeper route to the sea. If this happened it would bypass the ports of Baton Rouge and New Orleans and suddenly deprive them of the shipping facilities that handle 50 million tons of exports a year. While struggling to find a new course it would wash away roads, railways and gas pipelines supplying 28 eastern states, and flush coastal communities like the

50,000 inhabitants of Morgan City out into the Gulf of Mexico.

A disturbing study from Louisiana State University points to the present record flow of the Mississippi, with its volume up nearly 50% from that of 50 years ago. Raphael Kazmann, a civil engineering professor at Louisiana, believes there could be an 'awesome' flood 'within a year or two'.

Deforestation and Floods

We saw in an earlier chapter the devastation caused to the global ecology by deforestation and its contribution to the carbon dioxide problem. It is worth remembering that trees are nature's own flood barriers, too.

The dense vegetation of some floodplains has a spongelike quality about them. The dark humus held in place by the roots of plants are protected by their tops from the harshness of the elements. When plants die their dead leaves rot in place to form a mulch or blanket. Water landing on this percolates down to the soil beneath, which absorbs it without sinking in too quickly and saturating the ground. The worst that can happen, before the flood-stage is reached — is that the area will become a marshy quagmire. But in the absence of vegetation most torrential rainstorms will result in damaging floods as the water will simply be unable to percolate into the ground fast enough.

Not only that but de-vegetated or deforested land would fast become impermeable as soon as the binding action of plant roots is destroyed and the soft topsoil removed through erosion. It also suffers irreversible biological changes. What few minerals that may be left in the subsoil are washed deep down, where it becomes lost to whatever plant life struggles vainly to gain a foothold in more favourable conditions. And the Sun, now unobscured by a sheltering forest canopy, bakes the ground hard.

Thus the certainty of more floods, and the perpetuation of a vicious circle. The porosity of the ground is not only eroded by the wind, but by torrential rain. The topsoil soon gets washed away, often at the rate of several inches a year. Thus each succeeding deluge becomes that much more damaging. One experiment in the Amazon showed that 85 inches of rain a year removes less than half a ton of soil per acre from a forested area sloping at 15 degrees. But when an area of forest is removed, even though it is on a level terrace, 45 tons of soil is washed away. Already the interior of large continents have less rainfall than coastal regions, and the vegetation is thus much sparser. Yet when the rains come, they fall in torrents, and encounter the least resistance.

Once the vegetation has gone, or nearly so, the shape of the terrain then becomes an important variable. The rainfall in the Western Mississippi Valley is only about 15 inches a year, and the vegetation is short grass and shrub. But the hinterland is nearly 1,500 feet above sea level. So the floods fall from a greater elevation, thus adding to its weight-uprooting properties all the way on its southbound journey to Oklahoma.

Thus it is imperative that the right sort of greenery is planted. In his famous early doomwatch book, *The Road to Survival*, William Vogt wrote that the corn and soyabeans planted in row-tills were the wrong sort of crops for the state of Iowa, as they were more easily eroded than forest or grassland. Vogt was writing after the 1947 Missouri floods that tore away more than 115 million tons of rich loam topsoil, which had hitherto made Iowa one of the greatest agricultural areas in the world.

Deforestation, as we have seen, has undoubtedly contributed to China's fearsome flood problems. Dr. Hu Qingjun, a river control expert, has said that highly destructive floods occurred in the Sichuan Province in the 19th century when extravagant rulers plundered the hillsides for timber to build their palaces. What he did not mention was that the 'grain first' agricultural policy practised under Mao Tse-Tung encouraged inappropriate terracing and other landscaping practices that greatly increased erosion.

Rivers on the Rampage

Rivers are nature's own flood release mechanisms. Most of the great floods in history are the result of rivers failing to cope with an excess of storm water that has not been held back by the Earth's natural vegetative flood barriers. But each river that bursts its banks turns the ratchet one notch further. The overflowing river goes through a period of adjustment to its self-modified flow and to the volume and nature of the sediment transported. Even in its drier phase the river remains a potential threat to life and property because of this ratchet effect. Like a self-fulfilling prophecy the river in its floodstage further increases its power to erode the channel and valley sides, thus making the next overflow more disastrous than the one before.

The destructive potential of rivers, too, often arises from their extraordinary power to shift massive boulders further downstream. Occasionally the waterway is forced into a totally new route towards inhabited areas. This has been the case with China's Hwang Ho — 'China's Sorrow'. This mighty river, loaded with yellow sediment, has an unenviable reputation for causing the deaths of more innocent human beings than any other agent of natural disaster anywhere in the world. Its most dangerous characteristic is that it often runs well above the surrounding flat and fertile land.

In the past the Hwang Ho has claimed, and can still do so today, the lives of several thousand people during one overflow. And at intervals of about 150 years or so the death toll tops the million mark. It nearly did this in 1887, when some 900,000 died. In some places the river is a mile wide, and traverses a massive plain in which up to 100 million people live. Hubert Lamb, writing his mammoth *Climate: Past, Present and Future* (1982), says that Chinese river floods in 1332 accompanied by heavy rain, killed as many as seven million. This disastrous flood probably, claims Lamb, destroyed sanitation arrangements, encouraged the spread of plague-carrying rodents, and brought about the Black Death of the following year.

The Chinese today, however, as in other aspects of public policy, are now recanting of past errors. A programme of reafforestation is now being undertaken on the plains surrounding the Yangtze and Hwang Ho. It was only the low water level in the Donting Lake, which drains much of the Yangtze's water, that prevented the damaging floods of 1981 (which cost the country £700 million) from reaching the major cities of Wuhan and Shanghai.

Civilizations at Risk

Why, then, if rivers can be such killers, do people live in flood risk areas? The answer, of course, is invariably due to economic necessity. An estimated half a billion people live on floodplains, with untold millions more living along vulnerable coastal areas. Perhaps as many as 200 million Chinese live in the Hwang Ho and Yangtze River valleys.

Something like one-third of the world's population is fed from produce grown in fertile floodplains. Indeed, as famine spreads across those other regions plagued by drought the river valleys of the world seem more desirable places to live in. The soil is often deep and easy to cultivate, containing deposits of rich alluvium, highly beneficial for agriculture. The vast slimy sea of liquid mud surrounding river banks after a flood makes an ideal fertilizer.

Floods are really human problems, in the sense that the authentic history of floods in America is generally less than 200 years, after sizeable populations had become settled, compared with centuries in Europe and Asia. Floods only became noticeable in the States from the late 18th century onwards, while the West was being won, and migrants arrived from Europe to populate the east coast and the mid-West.

Nowadays, across the world, floodplains are being increasingly occupied by the infrastructure of civilization — homes, factories, highways and public utilities. As the population has grown, more buildings are erected, more public utilities are installed, more land is farmed and more food is eaten. As communications spread — rail, automobiles, telegraph wires — so disruption is that much greater when the floods come.

Not only is the flood problem human, but often the immense suffering and devastation caused by floods are themselves manmade. Because of poverty many victims in the under-developed nations die after their flimsily built houses collapse around them or are swept away. The inhabitants of poorer countries, by a tragic geographic irony, are more prone to earthquakes and fierce cyclonic storms than those living in the rich northern hemisphere. In Third World countries there are some 3,000 disastrous deaths every year, while in the West there are only about 500.

The risk of drowning or injury is made worse by the massive growth of illegal or semi-legal squatters living in makeshift property. According to the 1984 *Earthscan* publication on natural disasters referred to in Chapter Six, up to 75% of urban populations can live in swamplands or other dangerous areas. Often the land is totally without the solid structures, barriers and flood control mechanisms we take for granted in the developed world. For instance the *Earthscan* report claimed that in Guayaquil, Ecuador, some 60% of the population of over one million live in squatter communities built on catwalks over swampland. And along the steep slopes of Rio many flimsily built shanty town dwellings literally slide down and disintegrate when rainstorms loosen the soil. As many as 1½ million of Delhi's inhabitants live permanently in makeshift squats in the floodplains of the Yamuna River.

Apathy and fatalism compound the failures of Third World authorities to tackle the flood problem. John Whittow, in his book *Disasters*, cites the depressing case of the villagers on the Ganges floodplain in India. Very few were ignorant of the flood hazard, but devolved moral responsibility for flood management onto the state. They preferred to bear the loss to crops, livestock, and housing rather than move to another area.

The neglect of the authorities, of course, is because of the acute shortage of funds for planning and preventative measures. Even those governments, like the Chinese, who try to implement remedial measures, find their best efforts defeated. Since the war the Chinese have undertaken a great deal of flood alleviation work. In Honan Province, as well as replanting the forests, they are engaged in a comprehensive programme of dike raising. River flow velocities have been slackened by special piers, baffles and dams. But the build-up of sediment constantly vitiates the best efforts to control the course and flow of rivers. The Sanmen gorge dam opened in 1960 could originally control 92% of the water, but this capacity is already down to 50%.

Ironically, improved techniques of flood control in the West can actually increase the numbers at risk by instilling a false sense of security in the minds of homesteaders. According to Maurice Arnold, a director of the US Bureau of Outdoor Recreation, earlier populations at risk were encouraged to return to the floodplain. This turned out to be a dangerous practice. The risk of flooding was reduced, but far from eliminated, as much greatly depended on the vagaries of topography and climate.

People, furthermore, when confronted with the risk of drowning (as with the risk of crossing the road), looked only at the statistical probability of it occurring. Instead it would be wiser to check into local history, or look for evidence of extreme meteorological violence, as drawn to their attention frequently enough in the media. For example, in the Hunter Valley, in the coastal region of Australia's New South Wales, very few new residents were able to respond accurately to a questionnaire concerning the likelihood of floods, despite the lengthy record of serious inundations in the area.

Many more millions, however, will have no choice about where they live, or about the kind of risks that living in vulnerable topographical areas entails. We have seen how most urban areas have grown up around meandering river basins, and how the postwar trend has been to migrate to large cities, rather than away from them. Let us take one of the most heavily urbanized areas in the world — Europe — where rivers overflow with varying degrees of severity and with disturbing regularity. In Spring 1983 European skies wept endlessly. All of the major river systems of Europe lapped over their banks, some up to four times: the Seine, Garonne, Loire, Rhone, Rhine, Moselle, Neckar and Saar. The Rhine, in fact, swollen by torrential rain of long duration, roared in a brown torrent through the old city centre of Cologne, up to five feet deep. The Speaker's private entrance to the Bundestag building in Bonn could only be reached by boats.

The relocation of flood-affected people is always difficult. Cities, farms and businesses are all at risk. The effects of flooding as a result of climatic change, however, will be gradual, and will enable people and industries to find flood-risk-free sites.

There are glimmers of optimism. Scientists are confident that we have ways of successfully dealing with the floodwave problem. One obvious answer is to build more dikes, as the Dutch have had to do for centuries (and who have suffered far more from the real and damaging effects of sea encroachment than any other nation). Jim Titus recommends a major sea wall building programme in the US to protect vulnerable coastal areas. Natural sand and earth mounds can be hauled onto beaches to combat erosion. Zone ordinances could prohibit construction in lowlying tracts. Possibly reservoirs and catchment basins could be enlarged and strengthened to aid the regions more prone to desertification.

As the floodwave rolls inexorably nearer, many fertile coastal areas will have to be abandoned, and attempts to cultivate other areas further inland will be hampered by changes in climate, which may mean an overall decline in output from the present agriculturally productive areas of the world, and even food shortages.

In conclusion, therefore, we must take a brief look at which countries are likely to benefit, and which to lose, when the warming reaches its peak.

Epilogue

WHO WINS, WHO LOSES?

One fact stands out about the future warming: within about 200 years from now the world's existing fertile regions will be adversely affected as their climates grow more arid, and more and more arable and agricultural activities will be done in the less climatically favourable highland regions.

The general trend will be for farming to move towards the poles, and yields to become reduced. Many fertile regions will be eroded by both floods and droughts, and there will be the silting up of arable regions situated too near the world's coasts. The midwest corn belt of America might be replaced with a corn belt stretching across Saskatchewan. America's bumper crop could become a thing of the past. The northernmost land masses, including parts of Canada, Alaska and Scandinavia would all benefit. An increase in Iceland's grazing land could result in a doubling of its sheep production.

On the other hand precipitation levels would increase in many parts of Asia, including Russia. Her climate may become more equable, rather like that of Western Europe today. Russia's growing wealth, as she soon overcame her notorious food shortages, and actually became a net food exporter, could heighten global tensions and instability.

And yet parts of north and east Africa, the Middle East, Mexico and Australia, all of which are either permanently or periodically drought stricken, could, according to Roger Revelle, an oceanographer at the University of California, actually become grain exporters as well. India might be able to grow an extra crop of rice a year. The UK could soon have a climate like that of the Middle Ages, with flourishing vineyards.

Hitherto a long run of excellent crop-producing weather in the

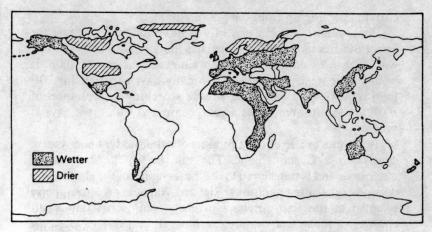

100 years from now scientists reckon the world's climate could already be showing changes on this scale.

Source: Will Kellogg, Climatic Change, (ed) John Gribbin, Cambridge University Press, 1978

United States since the war has induced a sense of complacency only recently shattered by weather aberrations. But now technological developments in farming oblige the farmer to reduce his variety of crop species; to become overly dependent on just one or two crops. He therefore becomes more vulnerable to prolonged or disturbed changes in the weather if his crops fail.

As time passes Herman Flohn, a distinguished German climatologist at Bonn University, says that, with US temperatures some 10F above the median, the water supply situation in California and Utah would become 'catastrophic'. Low lying areas on both sides of the Atlantic would suffer worsening storm damage. Many rivers like the Delaware would start to become saline, and marshes would become saltier as seawater migrated further up estuaries.

There would be distinct changes in the circulation patterns of the oceans, which could have a bearing on marine life. Bill Jenkins, a US oceanographer, believes the Gulf Stream and other currents could weaken because of the reduced thermal contrast between the poles and the Equator. The strength of the oceanic 'pump' would be weakened. If the Gulf Stream slowed, this could mean that Britain would no longer be bathed in the warm ocean currents coming up from the South Atlantic.

Climatologist Jim Hansen of NASA's Goddard Institute for Space Studies in New York calculates that by 2180 AD New York would have 12 days a year when temperatures exceed 100F. It would have nearly three times as many days experiencing 90F than now. Omaha would experience scorching temperatures of 100F for about three weeks a year, compared with three days at present.

Hurricanes are, in effect, a product of a thermodynamic system, similar to a Carno engine. The rate at which ocean water evaporates and is transferred to the 'heat engine' of the atmosphere depends on sea temperatures. Richard Warrick, an atmospheric scientist at the East Anglia University, points out that small changes in mean temperature can have an exaggerated effect on the world's weather systems, and increase the frequency of freezes, droughts and floods, and storms.

Similarly Kerry Emanuel, of the Massachusetts Institute of Technology, writing in a *Nature* article in April 1987, said rising ocean temperatures may increase the destructive power of tropical hurricanes by as much as 50%, particularly in the Gulf of Mexico and the Bay of Bengal. This is because the force of a cyclone varies according to the square of the wind speed. Dr Emmanuel's calculations show that a warm-up of, say 3C, will reduce the pressures at the centres of tropical cyclones considerably, sufficient to increase the average speed of winds blowing into the already low pressure regions at the heart of the cyclone by 15-20%.

Many of these predictions are arrived at by working backwards. A reconstruction by NCAR's William Kellogg says that in olden times the highlands of Mexico, and the northwestern part of the Indian sub-continent and north Africa, were wetter, whereas the central plains of North America were drier. Herman Flohn reviewed the warmer climates of the north as far back as the Late Tertiary, some 12 million years ago. From this he inferred that with a modern climate some 7F warmer, northern forests would even start growing near the North Pole. Paul Waggoner, director of the Connecticut Agricultural Experiment Station, has calculated that the gradual drying out of the central plains of America would reduce the flow of the Colorado River by 40%, and in the process deprive most western states of an adequate water supply.

As time passes it will be those countries that already have marginal lands, and in which the populations will have few opportunities for internal or overseas migration, that will suffer food shortages and privation. For the rest of the world the warming will demand vital changes in agricultural practices. New plants, adapted to cope with changing levels of atmospheric carbon dioxide and climatic shifts, would have to be bred. There will be a need for new methods to monitor and control both the CO^2 and the heat balance to keep Earth's temperature within manageable limits.

American farmers would probably be the best able to adapt in the short term. They are already experienced, through the exigencies of the free market, in switching to new crops in response to changing demands, or when existing crops start to become disease-prone. Farmers would seek to diversify their seed stocks, and become efficient in their use of water.

Governments, for their part, would need to implement radical re-afforestation programmes, and put pressure on Latin American countries to curb their tree-lopping practices. They would need to plan irrigation projects to take into account future shifts in rainfall patterns. There are broader international issues, such as making a much more determined effort to improve international relations, so that peaceful nations can be better able to mutually adapt trading arrangements.

Says George Woodwell, of the Marine Biological Laboratory: 'The series of major problems associated with carbon dioxide build-up will become major issues in the next century, whether we address them now or not'.

POSTSCRIPT

The encouraging news, as this book goes to press, is that some international progress is being made on controlling some of the environmental disbenefits that are adding to the Greenhouse Effect. Much of this seems to be directed towards enforcing an effective ozone-protection regime. Under pressure from the US, the Council for Ministers of the EEC decided in Brussels on March 19 1987 that they are ready to negotiate a global reduction of chlorocarbons. Even the Soviet Union and its allies, hitherto reluctant to get involved in environmental projects, has agreed to participate in ozone discussions.

However, the economic stakes are high. Literally thousands of companies use hydrocarbons on a large scale in the manufacture of cosmetics, electronics and domestic freezers, amounting to an annual value range of $2 billion. Even McDonald's hamburgers are sold in polystyrene cartons, and when they are crushed they release yet more CFCs into the atmosphere.

Some pressure groups are pushing for the development of CFC-22 and CFC-134a, both promising substitutes as refrigerator coolants. Britain's ICI and Germany's Hoesch AG have taken out patents on CFC-134a. Many big car manufacturers and electronics firms are searching for substitutes to the suspect chemicals. But other major European manufacturers are pleading that there is still insufficiently hard evidence that CFCs are the main ozone-destroyers, and have not sought to promote alternative products.

However, as we have seen, the greatest environmental threat comes from deforestation, if not from population growth itself. The twin problems are the wastefulness of the hunter-gatherer and the logging activities of commercial undertakings. The main problem is exploitation rather than harvesting.

There are glimpses of hope in Europe. Over-production of food has stirred the British government into encouraging farmers to take land out of cereal and dairy population. For the first time in the West, with the incentive of tax concessions and investment grants, it might be profitable to plant trees. May experts can show that forests can be harvested on a commercial scale (in the same way that fish can be 'farmed') far more easily than is often thought as is the case of the large newsprint industries in Scandinavia, where changes in husbandry have had to be made to ensure future supplies.

Unfortunately it will be a long time before Third World governments will be able to compensate the beef farmers of Brazil to enable them to follow the British example, or to provide cheap commercial sources of energy to the slash-and-burn peasants of Chad, Tanzania and Burundi. Recently the United Nations Agency for International Development said that for more than a third of the world's population firewood is still the primary source of energy. For another billion, wood fulfils more than half their fuel needs. And where reafforestation takes place in the Third World, it is never sufficient. India eliminates about a million hectares of forest a year, and replants less than half a million.

Nevertheless in 1987 — European Year of the Environment — there is clear evidence that international CO^2 monitoring bodies are doing more to publicize the Greenhouse issue. They are becoming more strident. One group, sponsored by the World Meteorological Organization, The United Nations Environment Programme and the International Council of Scientific Unions (ICSU), with scientists from 29 countries, declared that it is no longer good enough to assume a steady-state climate, and important social and economic decisions about agricultural land use, hydro-power and water projects etc. must take into account the effects of a gradually warming climate.

The time has probably come for an additional international conference on the greenhouse effect to iron out some of the scientifically disputed factors concerning sunspot activity, volcanism and wind patterns, as well as the other non-anthropogenic factors discussed in this book. This, admittedly, is a tall order, and probably explains why international concern is more easily focused on known human activities since this is the one variable that is within our control.

In the meantime, there must be renewed attempts to get consumers to be more efficient in their use of energy, and to eliminate fuel waste. Failing this, the human factors affecting the global warming will bring about irreversible climactic change, with the mixed benefits and disbenefits discussed in the Epilogue, within a hundred years from now, and much of this change could be harmful to the biosphere as we know and understand it. Sadly, many respected scientific bodies believe that it is already too late.

Bibliography

BOOKS

Allaby, Michael, and Bunyard, Peter, The Politics of Self-Sufficiency, Oxford University Press, 1980.
Allison, Ian (ed), Sea Level & Climatic Change, International Association of Hydrological Sciences, Canberra, 1979.
Asimov, Isaac, A Choice of Catastrophes, Hutchinson, 1979.
Asimov, Isaac, From Heaven to Earth, Dobson Books, 1968.
Asimov, Isaac, Change!, Coronet Books, 1983.
Asimov, Isaac, Please Explain, Coronet Books, 1982.
Bascom, W., Waves & Beaches, Anchor Press, 1980.
Bernard Harold W., The Greenhouse Effect, Ballinger (US), 1980.
Bloomfield, Arthur E., The Changing Climate, Dimension Books (US), 1977.
Booth, Basil, & Fitch, Frank, Earthshock, Dent, 1980.
Breuer, G., Air in Danger, Cambridge University Press, 1980.
Brubaker, Sterling, To Live on Earth, Mentor (US), 1972.
Calder, Nigel, The Weather Machine, BBC, 1974.
Canning, John, Great Disasters, Octopus Books, 1976.
Carr, Donald E., Energy & the Earth Machine, Abacus, 1978.
Claiborne, Robert, Climate, Man & History, Angus & Robertson, 1970.
Clube, Victor, & Napier, Bill, The Cosmic Serpent, Faber, 1982.
Coates, Donald R., Environmental Geology, J. Wiley & Sons, 1981.
Cornell, James, The Great International Disaster Book, Charles Scribner & Sons (US), 1976.
Critchfield, Howard J., General Climatology, Prentice-Hall (US), 1974.
Dacy, D.C. & Kunreuther, H., Economics of Natural Disasters, Collier-Macmillan (US), 1969.
Dasman, R.F., Planet in Peril, Penguin, 1972.
Ecologist, The, Blueprint for Survival, Penguin, 1972.
EPA, The, Projecting Future Sea Level Rise, Washington, 1983.
EPA, The, Can We Delay a Greenhouse Warming?, Washington, 1983.
Gaskell, T.F., & Harris, Martin, World Climate, Thames & Hudson, 1979.
Goudie, A.S., Environmental Change, Clarendon Press, 1977.

Grazier, Alfred de, The Velikovsky Affair, Abacus, 1978.
Gregory, K.J., & Walling, D.E. (eds), Man & Environmental Processes, Dawson Westview Press, 1979.
Gribbin, John, The Strangest Star, Fontana, 1980.
Gribbin, John, Carbon Dioxide, Climate & Man, Earthscan, 1981.
Gribbin, John, Future Weather, Pelican, 1983.
Gribbin, John, Genesis, Dent, 1980.
Gribbin, John & Plagemann, Stephen, Beyond the Jupiter Effect, Macdonald, 1983.
Halacy, D.S., Ice or Fire? Barnes & Noble (US), 1980.
Harvey, Brian, & Hallett, John D., Environment & Society, Macmillan, 1977.
Hassler, Gerd von, Lost Survivors of the Deluge, Signet (US), 1978.
Hatherton, T. (ed), Antarctica, Methuen, 1965.
Hewitt, Ronald, From Earthquake, Fire and Flood, Allen & Unwin, 1957.
Higgins, Ronald, The Seventh Enemy, Pan Books, 1980.
Hilary, Edmond (ed), Ecology 2000, Michael Joseph, 1984.
Hollis, G.E., Man's Impact on the Hydrological Cycle, Geo Abstracts, 1979.
Holt & Langbein, Floods (US), 1955.
Jager, Jill, Climate & Energy Systems, John Wiley, 1983.
King, M.G.R., The Antarctic, Blandford Press, 1969.
Lamb, H.H., Climate, History & the Modern World, Methuen, Methuen, 1982.
Landsberg, Helmut E., The Urban Climate, Academic Press, 1981.
McDonald, G.J. (ed), Long Term Impact of Increasing CO2, Ballinger (US), 1982.
McGraw, Eric, Population Today, Kaye & Ward, 1979.
McWhinnie, Mary A. (ed), Polar Research, Westview Press, 1978.
MIT, Inadvertent Climate Modification, MIT Press (US), 1971.
Moore, Patrick, Can You Speak Venusian?, Wyndham Publications, 1976.
Myers, Norman, The Sinking Ark, Pergamon, 1979.
Milne, Antony, London's Drowning, Thames Methuen, 1982.
Milne, Antony, Floodshock, Alan Sutton, 1986.
Neal, Philip, Acid Rain, Dryad Press, 1985.
North, Richard, The Real Cost, Chatto, 1986.
Olsen, Ralph E., Geography of Water, Wm C. Brown & Co (US), 1970.
Pittock, Frakes et al, Climatic Change & Variability, Cambridge University Press, 1978.
Ponte, Lowell, The Cooling, Prentice-Hall, (US), 1976.
Raikes, Robert, Water, Weather & Prehistory, John Baker, 1967.

Bibliography

Righter, Rosemary, & Wilsher, Peter, The Exploding Cities, Deutsch, 1975.
Schneider, Stephen, The Genesis Strategy, Plenum Press (US), 1976.
Sears, Paul B., Deserts on the March, Routledge, 1949.
Taylor, Gordon Rattray, The Doomsday Book, Thames & Hudson, 1970.
Taylor, Gordon Rattray, How to Avoid the Future, New English Library, 1977.
Velikovsky, Immanuel, Earth in Upheaval, Gollancz, 1956.
Vogt, W., Road to Survival, Gollancz, 1949.
Waltham, Tony, Catastrophe, Macmillan, 1978.
Walworth, Frank, Subdue the Earth, Panther, 1977.
Ward, Barbara, Progress for a Small Planet, Pelican, 1979.
Ward, Ron, Floods, Macmillan, 1978.
Warlow, Peter, The Reversing Earth, Dent, 1982.
Warshofsky, Fred, Doomsday, Sphere, 1977.
Weil, Robert (ed), Omni Future Almanac, Sidgwick & Jackson, 1983.
White, John Pole Shift, W.H. Allen, 1980.
Whittow, John, Disasters, Allen Lane, 1980.
Wijkman, Anders, & Timberlake, Lloyd, Natural Disasters, Earthscan, 1984.

NEWSPAPERS & JOURNALS

Attenborough, D., Observer, 1/4/84.
Berry, Adrian, Telegraph, 6/10/86.
Brandt, John, Sunday Times, 15/7/84.
Binyon, Michael, Times, 30/5/83.
Browne, Malcolme, Discover (US), February 1984.
Brummer, Alex, Guardian, 28/12/82.
Caulfield, Catherine, Guardian, 9/9/82.
Clayton, Hugh, Times, 1/10/86.
Davidson, Spencer, Time magazine, 13/6/83.
Edmunds, Tom, Time magazine, 6/1/84.
Goldfarb, Michael, Guardian, 17/11/86.
Gribbin, John, New Scientist, 15/3/84.
Gribbin, John, Science Digest (US), December 1982.
Gribbin, John, Guardian, 14/1/82, 15/8/86.
Hogg, Sarah, Times, 11/7/84.
Harding, Colin, Times, 13/7/83.
Jackman, Brian, Sunday Times, 4/4/82, Times 18/12/79.
Jensen, H.A.P., Nature Conservancy, 1953.
Johnson, Stanley, Observer, 5/8/84.

Jordan, Philip, Guardian, 7/4/80.
Hindley, Keith, Sunday Times, 15/8/82.
Lamb, Hubert, Times Higher Education Supplement, 2/12/83.
Lochenbruch, Arthur & Marshall, Vaughn, Science (US), 4/11/86.
McKean, Kevin, Discover (US), December 1983.
McKie, Robin, Observer, 29/1/84, 5/8/84.
Osman, Tony, Sunday Times, 13/11/83.
McCrea, W.H. Nature 255, 1975.
Milgram, Lionel, New Scientist, 19/5/85.
Pauley, Robin, Guardian, 9/2/80.
Pearce, Fred, New Scientist 18/9/87.
Prentice, Thompson, Times, 2/7/86.
Righter, Rosemary, Sunday Times, 5/8/84, 6/3/83.
Samstag, Tony, Observer 28/10/83.
Schwarz, Walter, Guardian, 1/7/83.
Silcock, Bryan, Sunday Times, 24/4/83, 4/1/81.
Tickell, Crispin, Times, 17/8/82.
Troitsky, Vsevolod, Novosti (USSR), vol 19, 1985.
Tucker, Anthony, Guardian, 24/12/83, 8/5/80, 25/7/86.
Wilkie, Tom, Independent, 6/11/86.
Wright, Pearce, Times, 19/6/84, 31/7/84.
Yulsman, Tom, Science Digest (US), February 1984.

OTHER SOURCES

Readers' Digest, Aug 1981.
Eros, vol 63, Aug 1982.
Nature, vol 289 12/2/82; vol 298 29/7/82; vol 297 19/5/82; vol 305, Sept 1983; vol 315, 314, 1985; vol 322, 1986; vol 326, 26/3/87.
Icarus, vol 50, 1982.
    Ambio, vol 13, June 1984.
    Times Higher Education Supplement 4/7/86.
    Times 1/10/86, 10/6/85.
    Eos, vol 66, 1985.
    WHO, 'Urban Air pollution 1973-80', Nov 1984.
    New Scientist 15/5/86, 16/10/86.
    Sunday Times 13/7/86.
    Time magazine 3/11/86.
A Matter of Degrees, World Resources Institute, Washington, 1987.

Index

Acid rain 52-7
Africa, Africans 10-15, 32, 45, 49, 50, 69, 75-7, 101, 119, 129, 140, 142, 145
Allaby, Michael 89
Alvarez, W & L 89
American Nat. Sc. Found. 110
Armagh Observatory 92
Arnold, F 80
Arnold, Maurice 138
Arrhenius, Svante 64
Asia, Asians 10, 11, 13, 15, 16, 45-51, 101, 114, 117-20, 123, 129, 131
Asian Inst. of Technology 119
atmospheric gases 36-43, 52, 54-6, 58-69, 82, 90, 105-6
Australian Nat. Sc. Found. 110

Bannerjee, Subir 92
Belgian Space Aeronomy Inst. 40
Bernard, W. 86
Blair, Thomas 18-19
Booth, Basil 122, 127
Bosnywash 18, 28
Brasseur, I. 40
Br. Antarctica Survey 42
Br. Assoc. 80, 82
Brimblecombe, Peter 54
Brown, H.A. 102
Bryson, Reid 77
Buhrke, T. 80
Bureau of Recreation (US) 138

Cameron, Richard 110
Caputo, Michael 116
carbon dioxide, carbon compounds 2, 34, 36, 37, 39, 40, 43, 44, 58-69, 82, 85, 105, 132, 143-5
Carter, Jimmy 56
Cassini, Jean 27
CEGB 57

Census Bureau (US) 10
Channel, English 1, 18, 120, 121
Chen, Robert 127
China, Chinese 10, 11, 14, 88, 102, 111, 117, 118, 123, 129, 131, 135, 136, 138
CIA 6
Cicerone, Ralph 105
cities, urbanization 10-21, 23, 25, 28, 30, 32, 42, 82, 118-20, 129, 130, 137, 138
Cleland, John 13
climate 6, 21-3, 38, 60, 81, 113, 130, 132, 139, 140-3, 146
Climatology Unit, Univ. of E. Anglia 3, 54, 78, 80, 82, 83, 85, 142
Clube, Victor 89
Coast & Geodetic Survey (US) 124
Cold Regions Research Lab. 142
Connecticut Agric. Expl. Stn. 142
Cornell, James 133

Dansgaard, Wilf 99
Davies, Paul 91
deforestation etc. 24, 45-57, 58, 67-8, 76, 84, 134, 135, 143-4
Dept. of Energy (UK) 57
Dept. of Energy (US) 83
Dept. of Science (Austl.) 110
Dept. of Environment (US) 36
Detweller, Jean 27
Du Pont 44

Eckholm, Eric 56
ecosphere etc. 5, 11, 22, 23, 30-3, 52, 59, 76, 129, 134, 146
EEC 3, 43, 44, 65, 144
Emmanuel, Kerry 142
Emiliani, Cesare 99, 109
Environment Prog. (US) 44

EPA 1, 2, 5, 32, 64, 69, 83, 85, 108, 124, 130
Europe 11-20, 26, 33, 34, 42, 44-9, 54-7, 65, 68, 72, 73, 78, 83, 85, 101, 106, 117-124, 130, 137, 139, 140, 144, 145

Farman, Joe 42, 43
Farmer, Graham 83
Finch, Frank 122, 127
Fletcher, Joseph 107
Flint, Richard 7
Flohn, Hermann 141, 142
floods 129-143
Flood and Agric. Comm. 45
Forest Service (US) 50, 51
Fornos, Werner 10
Freon 39, 44

Geol. Survey (US) 108
Geophysical Fluid Dynamics Lab. 68
geotechnics etc. 5, 60, 102-3, 109, 110, 112, 114, 116, 117
Gilliland, Ronald 80, 95
Goddard Space Flight Centre 42, 65, 142
Gold, Thomas 103
Gottman, Jean 17
Gow, Anthony 108-9
greenhouse effect 4, 5, 58-69, 72, 79-82, 85, 95, 99, 108, 144, 145
Gribbin, John 65, 82, 83, 93-4

Hansen, James 65, 142
Hays, Jim 97, 99
Hays, Paul 60
Heath, Donald 42, 43
Hoffman, John 69, 73, 85
Hollins, John 109
Howard, Luke 25
Hoyle, Sir Fred 90
Hutton, James 3

Ice Ages 2, 4, 64-5, 90-2, 94, 96, 99, 108, 112-6
ice caps, polar regions 2, 3, 6, 7, 23, 35-6, 40, 42, 66, 69, 84, 91, 99-101, 112-4, 127
Imbrie, John 97

India, Indians 10, 11, 14, 15, 33, 101, 132, 138, 142
Inst. of Ecology, Munich 54
Int. Council of Sc. Unions 145

Jenkins, Bill 141
jet streams 43

Karl, Thomas 74
Kazmann, Raphael 134
Keeling, David Charles 64, 67-9
Kelley, John 107
Kellogg, William 112, 142
Kelly, Mick 80
Kohl, Helmut 56

Lamb, Hubert 8, 78-9, 114, 136
Lamond Doherty Geol. Observ. 62
Landsberg, Helmut 25-6
Laplace, Marquis de 88
Latin America 11, 15-6, 33, 42, 45, 49, 51, 76, 77, 100, 101, 104, 117, 123, 131, 134, 137, 140, 142-3, 145
Lawrence Livermore Lab. 66
Leatherman, Stephen 127
Leibniz, Gottfried 3
Lochenbruch, Arthur 84
London 1, 14-6, 18, 24-5, 27, 120-22, 127, 132
Lovelock, J. 89

Mahlman, Jerry 68
Marine Biol. Lab. 4, 143
Marshall, Vaughn 84
Max Planck Inst. 80
McCrea, W.H. 92
McDonald, G.J. 110
McGraw, Eric 20
McMurdo Stn. 42
Mediterranean 5, 34, 47, 116
Meier, Mark 108
Mercer, John 112-3
meteors, comets etc. 2, 88-90, 117
Middle East 33-4, 45-6, 48, 50, 69, 140
Mintzer, Irwin 64
MIT 43, 142
Milankovitch 96, 99, 102, 112
Mitchell, Murray 33

Index

Napier, Bill 89
NASA 40, 42, 44, 68
Nat. Academy of Sc. 1, 3, 40, 64-6, 69, 122, 130
Nat. Centre for Atmos. Research 95, 112, 127, 142
Nat. Climatic Data Centre (US) 74
Nat. Forest Reserve (US) 51
Nat. Met. Service, Argentina 42
Nat. Research Council (UK) 85, 128
New York 15-17, 73, 123
North, Richard 24
Norwegian Inst. for Water Research 55
Nutalaya, Prinya 119

Oden, Svante 55
OECD 34
Office for Foreign Disaster Assist. (US) 129
Opik Ernst 92
Overrein, Lars 55
ozone 4, 36-44, 68, 90-1, 144

Physics Inst. Berne 106
Plagemann 93
pollution etc. 32-44, 52, 54-5, 59, 78
population etc. 10-23, 30, 50, 129-30, 137, 143-4
Population Concern 20
Population Inst. 10
Ponte, Lowell 80
Ptolemy, Claudius 101
Pythagorus 101

Qinjin, Hu 135

Revelle, Roger 128, 140
Righter, Rosemary 15
rivers 1, 48-9, 132-5, 136, 138, 142
Robinson, Elmer 104-5

sea level, oceans etc. 1, 4, 38-9, 60-4, 67-8, 81, 84, 89, 100, 104, 106-7, 110-28, 130-2, 139, 141
Segnastum, Mats 55
Shabtaie, Sion 108, 110
Shackleton, Michael 97
Schiffer, Robert 68

Schneider, Stephen 36, 80, 82, 87, 113, 127
Scripps Inst. 64, 67, 81
Singh, Jyoti 11
Solar Mesosphere Explorer 40
Somerville, Richard 81
Soviet Academy of Sciences 30
Stauffer, B. 106
Stedman, D.H. 54
sun, solar radiation 3-6, 22-3, 36, 38, 42-3, 52, 60, 62-3, 66, 68, 78, 86-99, 104, 107, 109, 145
Swedish Forestry Comm. 55
Swedish Soc. for Conserv. 55

Thames Barrier 1, 122, 125
thermal pollution, energy use 22-32, 38, 58, 145
tidal waves 7, 111-128
Timberlake, Lloyd 75
Titus, Jim 124, 139
Tokyo 15-6, 26-7
Toynbee, Arnold 5
Troitsky, Vsevolod 30-1
Turco, Richard 90

UN bodies 10-1, 15-6, 19-20, 30, 44, 51, 56, 145
Univ. of Berkeley, Calif. 89
Univ. of Bonn 141
Univ. of California 140
Univ. of Columbia 97
Univ. of Colorado 109
Univ. of Kansas 94
Univ. of Louisiana State 11
Univ. of Miami 77, 99
Univ. of Michigan 54, 60
Univ. of Minnesota 92
Univ. of New York State 42
Univ. of Ohio State 112
Univ. of Oxford 15
Univ. of Princeton 68
Univ. of Rome 116
Univ. of Sussex 92
Univ. of Washington 108
Univ. of Wisconsin 77
Univ. of Wyoming 42
Univ. of Yale 7
USA 7, 11-5, 18, 20-5, 33-4, 38, 44-5, 48-50, 54-6, 58, 69, 72-73,

83, 85, 102, 109, 112, 116-8, 123-44
USSR 11, 36, 42, 68, 84-5, 90, 123, 140, 144

Velikovsky, Emmanuel 88-9
Vogt, William 135
volcanoes 33, 43, 60, 62, 67, 77-80, 90, 145

Waggoner, Paul 142
Walker, James 60
Warlow, Peter 102
Warrick, R. 142
Watson, Robert 44
weather 5, 28-9, 35, 37, 66, 72-86, 94, 114, 118, 130-33, 137-143
West, Richard 82
Wigley, Tom 80
Wijkman, Anders 75
Wilsher, Peter 15
Whiston, W. 88
White, John 102
WHO 33
Woodwell, George 4, 143
Woodward, John 3
World Bank 21
World Resources Inst. 64

Young, N.W. 110

Zeller, Edward 94